Lecture Notes of the Institute for Computer Sciences, Social Informatics and Telecommunications Engineering 245

More information about this series at http://www.springer.com/series/8197

Peter Han Joo Chong · Boon-Chong Seet
Michael Chai · Saeed Ur Rehman (Eds.)

Smart Grid and Innovative Frontiers in Telecommunications

Third International Conference, SmartGIFT 2018
Auckland, New Zealand, April 23–24, 2018
Proceedings

 Springer

Editors
Peter Han Joo Chong
Auckland University of Technology
Auckland
New Zealand

Boon-Chong Seet
Auckland University of Technology
Auckland
New Zealand

Michael Chai
School of Electronic Engineering
  and Computer Science
Queen Mary University of London
London
UK

Saeed Ur Rehman
Auckland City Hospital
Auckland
New Zealand

ISSN 1867-8211   ISSN 1867-822X   (electronic)
Lecture Notes of the Institute for Computer Sciences, Social Informatics
and Telecommunications Engineering
ISBN 978-3-319-94964-2   ISBN 978-3-319-94965-9   (eBook)
https://doi.org/10.1007/978-3-319-94965-9

Library of Congress Control Number: 2018947452

Printed on acid-free paper

This Springer imprint is published by the registered company Springer International Publishing AG
part of Springer Nature
The registered company address is: Gewerbestrasse 11, 6330 Cham, Switzerland

Preface

We are delighted to introduce the proceedings of the third edition of the 2018 European Alliance for Innovation (EAI) International Conference on Smart Grid and Innovative Frontiers in Telecommunications (SmartGIFT 2018). Different from the last two SmartGIFT conferences that were held in the UK, this year, we brought SmartGIFT from the northern hemisphere to the southern hemisphere and it was held at Auckland, New Zealand, during April 23–24, 2018. SmartGIFT 2018 was co-organized by EAI and Auckland University of Technology. The theme of SmartGIFT 2018 was "Smart Grid and Innovative Frontiers in Telecommunications." This conference brings together researchers, developers, and practitioners from around the world who are leveraging and developing next-generation smart grids for a smarter and more resilient grid, and for advancing telecommunications as an important enabler for human-to-human, human-to-machine, and machine-to-machine connectivity.

The technical program of SmartGIFT 2018 consisted of 28 full papers, including five invited papers in oral presentation sessions. Aside from the high-quality technical paper presentations, the technical program also featured three keynote speeches and three invited talks. The keynote speeches were by Professor Soung Chang Liew from The Chinese University of Hong Kong, Hong Kong, Professor Tek Tjing Lie, from Auckland University of Technology, New Zealand, and Professor Maode Ma from Nanyang Technological University, Singapore. The three invited speakers were Professor Ho-Pui Ho from The Chinese University of Hong Kong, Hong Kong, Professor Yong Liang Guan from Nanyang Technological University, Singapore, and Professor Ivan Wang-Hei Ho from The Hong Kong Polytechnic University, Hong Kong. Selected best papers were invited to submit manuscripts to special issues of Springer's *Wireless Networks* on "Caching for Wireless Communication Systems and Networks" and AIMS *Electronics and Electrical Engineering* on "Mobile and Wireless Technologies for Sustainable Mobility and Transportation System."

Coordination with the steering chairs, Imrich Chlamtac, Bruno Kesler, Victor C. M. Leung, and Kun Yang, was essential for the success of the conference. We sincerely appreciate their constant support and guidance. It was also a great pleasure to work with such an excellent Organizing Committee team for their hard work in organizing and supporting the conference. In particular, we thank the Technical Program Committee, led by co-chairs, Dr. Xuejun Li and Dr. Faraz Hasan, who completed the peer-review process of technical papers and compiled a high-quality technical program. Other Organizing Committee members who supported and co-organized this conference were Dr. Ke Wang, Dr. Ramon Zamora, Dr. Saeed Rehman, and Dr. Hakilo Sabit. We are also grateful to the EAI conference manager, Lenka Bilska, for her support and all the authors who submitted to and presented their papers at the SmartGIFT 2018 conference. We strongly believe that the SmartGIFT conference provides a good forum for all researchers, developers, and practitioners to discuss all aspects of science and technology relevant to smart grids. We also expect that future SmartGIFT conferences will be as successful and stimulating as indicated by the contributions presented in this volume.

Peter Chong
Boon-Chong Seet
Michael Chai
Saeed Ur Rehman

Organization

Steering Committee

Steering Committee Chair

Imrich Chlamtac | Bruno Kessler Professor, University of Trento, Italy /EAI

Steering Committee

Victor C. M. Leung | The University of British Columbia, Canada
Kun Yang | University of Essex, UK

Organizing Committee

General Chair

Peter Chong | Auckland University of Technology, New Zealand

General Co-chairs

Boon-Chong Seet | Auckland University of Technology, New Zealand
Michael Chai | NZ Queen Mary University of London, UK

Technical Program Committee Co-chairs

Xuejun Li | Auckland University of Technology, New Zealand
Faraz Hasan | NZ Massey University, New Zealand

Web Chair

Ke Wang | Beijing University of Posts and Telecommunications, China

Publicity and Social Media Chair

Kin Kee Chow | Manchester Metropolitan University, UK

Workshops Chair

Maode Ma | Nanyang Technological University, Singapore

Sponsorship and Exhibits Chair

Ramon Zamora | Auckland University of Technology, New Zealand

Publications Chair

Saeed Rehman   Auckland University of Technology, New Zealand

Posters and PhD Track Chair

Minglong Zhang   Auckland University of Technology, New Zealand

Local Chair

Hakilo Sabit   Auckland University of Technology, New Zealand

Conference Manager

Lenka Bilska   EAI (European Alliance for Innovation)

Technical Program Committee

Stephan Cejka   Siemens
Faraz Hasan   Massey University, New Zealand
Peng-Yong Kong   Khalifa University, United Arab Emirates
Tuan Anh Le   Middlesex University, UK
Shunbo Lei   The University of Hong Kong, SAR China
Hui Lin   Fujian Normal University, UK
Keivan Navaie   Lancaster University, UK
Harold Chamorro   KTH Royal Institute of Technology, Sweden
Haozhe Wang   University of Exeter, UK
Chunbo Luo   University of Exeter, UK
G. C. Deepak   Lancaster University, UK
Subhonmesh Bose   University of Illinois at Urbana Champaign, USA
Khan Ferdous Wahid   Airbus Group Innovations
Zhibo Pang   ABB Corporate Research
Haijun Zhang   Beijing University of Chemical Technology, China
Maysam Qadrdan   Cardiff University, UK
Xiang Gui   Massey University, New Zealand
Mohammad Rashid   Massey University, New Zealand
Fakhrul Alam   Massey University, New Zealand
Et Lau   Queen Mary University of London, UK
Abbas Malik   Auckland University of Technology, New Zealand
Xiaobing Wu   University of Canterbury, New Zealand
Aziz Ahmad   Unitec Institute of Technology, New Zealand
Hnin Yu Shwe   Nanyang Technological University, Singapore
David Tung Chong Wong   I2R, Singapore
Yi Cen   Minzu University of China, China
Tuan Phung-Duc   University of Tsukuba, Japan
Lu Lu   Chinese Academy of Science, China

Contents

Temporary Internet Access
for Authentication and Key Agreement
for LTE Networks

Xue Jun Li[1]([✉]), Maode Ma[2], and Jiecheng Xie[2]

[1] Department of EEE, Auckland University of Technology, Auckland, New Zealand
xuejun.li@aut.ac.nz
[2] School of EEE, Nanyang Technological University, Singapore, Singapore
{emdma,jcxie}@ntu.edu.sg

Abstract. Evolved Packet System-Authentication and Key Agreement
(EPS-AKA) is the security protocol in Long-Term Evolution (LTE).
However, it is still vulnerable to user identity attacks and fake eNBs. Efficient EPS-AKA (EEPS-AKA) was proposed with some improvements.
Nevertheless, the EEPS-AKA is vulnerable to denial-of-service (DoS)
attacks and fake eNBs, despite of some minor flaws in its procedures. In
this paper, we propose Temporary Internet Access (TIA)-AKA to: (1)
prevent user identity disclosure by implementing some additional steps,
which allows a user equipment (UE) to request a temporary UE identity to access Internet; and (2) authenticate the Mobility Management
Entity (MME) through the validity of the assigned IP address. Physical
address and simple password exponential key exchange (SPEKE) method
are combined into the proposed TIA-AKA. Efficiency analysis suggests
the TIA-AKA provides a fully protection on the user identity and prevent the DoS attack, at the expense of increased bandwidth consumption
and processing delay.

Keywords: Wireless communications · Long Term Evolution
Security attack · DoS attack · Authentication and Key Agreement

1 Introduction

Long Term Evolution (LTE) was proposed to support high data rate, low latency,
multimedia traffic for future generation of cellular networks [9]. As shown in
Fig. 1, a LTE network consists of the Evolved Universal Terrestrial Radio Access
Network (E-UTRAN) and the Evolved Packet Core (EPC). E-UTRAN includes
evolved nodes B (eNBs), which communicate with user equipments (UEs). The
EPC is a fully packet-switched backbone network in LTE. Voice service will be
handled by the IP Multimedia Subsystem (IMS) network. The EPC consists of
a Mobility Management Entity (MME) and Serving Gateway (SGW), a Packet
Data Network (PDN) gateway together with Home Subscriber Server (HSS).
When a UE connects to the EPC, the MME performs a mutual authentication

© ICST Institute for Computer Sciences, Social Informatics and Telecommunications Engineering 2018
P. H. J. Chong et al. (Eds.): SmartGIFT 2018, LNICST 245, pp. 1–10, 2018.
https://doi.org/10.1007/978-3-319-94965-9_1

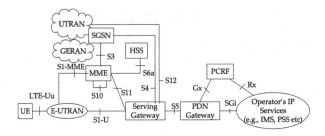

Fig. 1. LTE architecture of E-UTRAN and EPC [13]

with the UE [6]. The SGW forwards user data packets. The PDN GW allows a UE to connect to external packet data networks and allocates IP address to the UE [12]. With the fast deployment of 4G LTE networks, their vulnerabilities to certain security attacks attracted significant research interests.

In this paper, we study existing security schemes in LTE and propose a new scheme to improve the performance. Section 2 presents the related work, followed by the review of EEPS-AKA in Sect. 3. Section 4 presents our proposed scheme and Sect. 5 discusses its formal verification and performance evaluation. Section 6 concludes the paper with possible future work.

2   Related Work

The current security protocol adopted in LTE is Evolved Packet System-Authentication and Key Agreement (EPS-AKA) [15], which evolved from 2G-AKA, 3G-AKA to its current form. However, it is still vulnerable to user identity attacks and fake eNBs [3,8,19]. A UE will perform a new registration every time it connects to a new MME due to the fact that the new MME cannot obtain the UE's old Globally Unique Temporary Identifier (GUTI) to retrieve its International Mobile Subscriber Identity (IMSI). The user identity can be revealed when the IMSI is sent in plaintext during the registration process, which allows user identity attack. Similarly, IMSI may be sent to the fake eNB if it acts as a new MME by blocking the signal of real eNBs.

Several solutions were proposed to address these drawbacks. A comprehensive survey of existing researches and studies of LTE and LTE-A networks on security aspects was presented in [6]. In [7], EAP-FAKA was proposed to reduce the authentication delay and signaling cost. However, EAP-FAKA is vulnerable to fake eNBs [17]. In [10], I-AKA with GUTI was proposed to prevent DoS attacks. However, it cannot protect user identity when a UE registers for the first time. In [16], SE-EPS-AKA was proposed based on Wireless Public Key Infrastructure (WPKI), which suggests that UE, MME and HSS shall acquire the digital certificate via Certification Agency (CA) before communication. In [1], a new modified attack "Intelligent brute force" was presented. Nevertheless, it did not explain how an intruder knows the algorithm and the user identity is still vulnerable.

In [11], a HSK-AKA was proposed where digital signature is used to prevent malicious MME. However, the IMSI was encrypted with the new secret key and HSS required to know the IMSI in order to retrieve LTE key and calculate this secret key. Therefore, a contradiction occurs when a HSS cannot retrieve the IMSI, implying that it cannot read the messages from a UE at all. In [5], a solution was proposed to prevent DoS attacks that UE is required to attach its physical address to the authentication request.

The aforementioned protocols focus on securing the IMSI between UE and MME. Nevertheless, none of them aim to authenticate the authenticity of the MME. Thus, a fake eNB can still request the IMSI from UE as long as same protocol is used. In [2], MEPS-AKA was proposed based on the SPEKE method to provide strong mutual authentication between UE and MME, however, it cost more execution time.

3   Review of EEPS-AKA

3.1   Analysis of EEPS-AKA

Efficient EPS-AKA (EEPS-AKA) was proposed to deal with the issue of user identity disclosure [4]. In EEPS-AKA, Extensible Authentication Protocol (EAP)-SPEKE is based on password shared only between peer and authenticator. It is resistant to both active and passive attacks such as man-in-the-middle (MitM), replay, password sniffing and brute force. It generates a strong session key that can be used in data encryption. The password can be saved in a safe manner. Figure 2 illustrates the mechanism of SPEKE protocol.

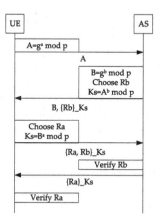

Fig. 2. SPEKE method

In EEPS-AKA, two random values (u and d) are chosen by UE to generate the key A, which makes the shared key always different even though same values (A^u, B^m) are used. The protocol starts when MME computes its value B and sends it to UE with user identity request message. After that, UE computers its value A using two random values (u and d), and the shared secret key K_{um} using f function, this key is used to protect the IMSI. When MME receives the protected IMSI (PIMSI), it calculates the K_{um} key and forwards it to HSS with other values. HSS and UE can verify each other via the random values computed by K_{um} and K_{uh} keys. To provide perfect forward secrecy, the secret key is used also to compute the generated keys in the later steps such as (IK, CK, and MSK). The details of the EEPS-AKA is in Fig. 3.

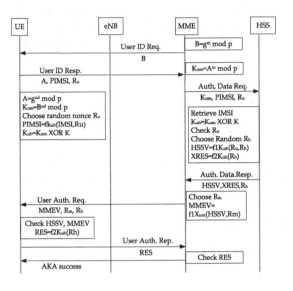

Fig. 3. Overview of EEPS-AKA

3.2   Flaws in the EEPS-AKA

Firstly, two of its computational algorithms the key generation process of EEPS-AKA, $A = g^{ud} \bmod p$ and $K_{um} = B^u \bmod p$, have problems in generating the secret keys. If these expressions are directly used to compute the secret key, it can be derived that:

$$K_{um} = A^m \bmod p = \left(g^{ud} \bmod p\right)^m \bmod p = g^{udm} \bmod p \qquad (1)$$

$$K'_{um} = B^u \bmod p = \left(g^m \bmod p\right)^u \bmod p = g^{um} \bmod p \qquad (2)$$

Obviously, K_{um} is different from K'_{um}. The proposed protocol cannot generate common session keys for UE and MME to communicate with each other, thus the MME cannot decrypt new messages received from UE. The mutual authentication between UE and MME fails at the very beginning. However, further analysis shows that the SPEKE is applicable in generating the session keys.

Assume random values u, d in UE and m in MME, the following algorithms can be used to generate the session keys:

$$K_{um} = A^m \bmod p = \left(g^{ud} \bmod p\right)^m \bmod p = g^{udm} \bmod p \qquad (3)$$

$$K'_{um} = B^{ud} \bmod p = (g^m \bmod p)^{ud} \bmod p = g^{udm} \bmod p \qquad (4)$$

Even though the idea is proved to be feasible, using two random variables in UE makes no sense in improving the SPEKE method. It may even result in wrong session keys if the g^{ud} or g^{udm} is larger than p as the previous assumption is valid only when g^{ud} or g^{udm} is smaller than p according to MATLAB simulations.

Secondly, the EEPS-AKA can generate strong session keys between UE and MME, which means that the messages between UE and MME cannot be revealed. Nevertheless, EEPS-AKA is still vulnerable to fake eNBs. The EEPS-AKA focuses on the protection of user identity rather than preventing DoS attacks, which can be launched with legitimate UEs. However, the detailed efficiency evaluation and simulation is not performed for EEPS-AKA in [4].

4   Proposed TIA-AKA Scheme

4.1   Motivation of TIA-AKA

Currently, there is no perfect solution to the problem of fake eNBs in LTE. It motivates us to propose Temporary Internet Access Authentication and Key Agreement (TIA-AKA) protocol, which utilizes the IP allocation scheme in LTE to distinguish fake eNBs.

4.2   Proposed TIA-AKA

TIA-AKA is based on EPS-AKA protocol and the SPEKE method used in EEPS-AKA. In addition, a special server is proposed to enhance the verification mechanism. TIA-AKA features a new mechanism for the UE to identify fake eNBs. UE requests temporary user identity for Internet access to check the authenticity of the MME/eNBs. It also combines the SPEKE method with MAC address to protect IMSI and prevent DoS attacks. As shown in Fig. 4, there are two sections and totally 10 steps for the TIA-AKA protocol. The first section is to validate the MME and the second section is mutual authentication among UE, MME and HSS. The 10 steps are:

(1) UE generates random variable u, computes $A = g^u \bmod p$, and sends the authentication request with A and its MAC address to MME via eNB.

(2) When MME receives the authentication request, MME records the MAC address and compares with its memory to avoid DoS attack. If the MAC address is fresh, the MME generates random variable m, computes $B = g^m \bmod p$ and uses the received A to compute the symmetric shared key $K_{um} = A^m \bmod p$. Then, MME sends a temporary identity request to the HHS.

(3) Upon receiving the temporary identity request, HSS generates a temporary identity request, authorizes the identity to be available for around 10 s and sends it back to the MME.

(4) The MME encrypts the temporary UE identity with K_{um} and send the message B, $\{TUI\}_{K_{um}}$ to UE.

(5) When UE receives the message B and $\{TUI\}_{K_{um}}$, UE computes symmetric shared key $K_{um} = B^u \bmod p$. UE uses the computed key K_{um} to decrypt the TUI and apply the TUI to request connection on P-GW. If the UE can get a valid IP address and connect to the Internet, the UE confirms that the MME is legitimate MME. *Furthermore, a server of service provider is specially set up for temporary identity authentication.* A special message and expected response with a special symmetric key is stored in SIM card when produced. Once UE accesses the Internet for authentication, it sends the special message with its temporary identity to the server. If the response tells that the identity is legitimate, UE confirms that MME is legitimate. Then, UE sends its IMSI, registration request and MAC address encrypted with the K_{um} to the MME.

(6) MME compares the MAC address again to prevent DoS attack and forwards the IMSI, SNID and n to HSS if the request is fresh.

(7) Upon receiving the authentication request from the MME, the HSS first verifies the IMSI and SNID and uses the retrieved LTE key and generated random RAND and SQN to create XRES, AUTN, CK and IK. Then, a top-level key (K_{ASME}) is calculated through Key Derivation Function (KDF) with the SNID, CK and IK. The HSS forms n AVs and sends them back to the MME. The $AV_i = (\text{RAND}_i, \text{AUTH}_i, \text{XRES}_i, K_{ASME_i}), i = 0, 1, \cdots, n$. MAC $= f_{1k}(\text{SQN}\|\text{RAND}\|\text{AMF})$, XRES $= f_{2k}(\text{RAND})$, CK $= f_{3k}(\text{RAND})$, IK $= f_{3k}(\text{RAND})$, AK $= f_{3k}(\text{RAND})$ $K_{ASME} =$ KDF $(\text{SQN} \oplus \text{AK}, \text{SN}, \text{id}, \text{CK}, \text{IK})$, AUTN $= \text{SQN} \oplus \text{AK}\|\text{AMF}\|\text{MAC}$.

(8) The MME stores the AVs received from the HSS, and selects one of them to use in LTE authentication of the UE. The MME allocates KSI_{ASME}, an index of K_{ASME}, and delivers it instead of K_{ASME} to the UE so that the UE and the MME can use it as a substitute for K_{ASME}. The MME sends KSI_{ASME_i} together with $RAND_i$ and $AUTH_i$ in the Authentication Request to the UE.

(9) Upon receiving the Authentication Request from the MME, the UE extracts the messages from the AUTH to check the received messages with following operations: XAK $= f_{5k}(\text{RAND})$, SQN $= \text{XAK} \oplus \text{SQN} \oplus \text{AK}$, XMAC $= f_{1k}(\text{SQN}\|\text{RAND}\|\text{AMF}) =?\text{MAC}$, XSQN $=?\text{SQN}$. If one of the two checks fail, it delivers Authentication Failure (CAUSE) message; otherwise, it calculates RES $= \text{f2k}(\text{RAND})$ and sends Authentication Response with RES back to MME.

(10) Once the MME receives the RES from the UE, it compares the RES with the XRESi of the AV received from the HSS. If RES matches the XRESi, the MME send a success message to UE and the authentication process is completed.

After completion of authentication, the UE derives K_{ASME} with CK, IK, SQN and SN ID. KSI_{ASME} received from the MME is used to represent the index of K_{ASME} and KSI_{ASME} is used during the NAS security setup between the UE and the MME. Note that these procedures are only processed when the UE registers to the MME and HSS for the first time; after success of the registration, GUTI is used instead of IMSI for other authentication process.

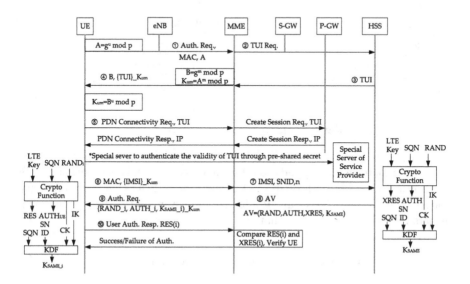

Fig. 4. Overview of proposed TIA-AKA

5 Security Analysis and Performance Evaluation

5.1 Formal Verification of TIA-AKA

The automated validation of Internet security protocols and applications (AVISPA) tool is used for validating the protocols. The AVISPA verification outputs of EEPS-AKA and TIA-AKA are shown in Fig. 5(a) and (b), respectively. From the execution outputs, we can see that the TIA-AKA is safe and it achieves the specified goals.

5.2 Performance Evaluation of TIA-AKA

Table 1 summarizes the length of authentication parameters [14]. For EPS-AKA, the bandwidth requirement [18] is given by

$$BW_{EPS\text{-}AKA} = (963 + 608n)\, N_{avg,AEPH} \tag{5}$$

where $N_{avg,AEPH}$ is the average number of authentication event per HSS. Similarly, for the proposed TIA-AKA, the bandwidth requirement is given by

$$BW_{TIA\text{-}AKA} = (1510 + 608n)\, N_{avg,AEPH} + (393)\, N_{avg,AEPM} \tag{6}$$

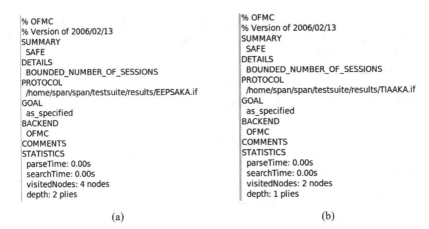

Fig. 5. AVISPA output of EEPS-AKA

where $N_{avg,AEPM}$ is the average number of authentication event per MME. In Eq. (6), the last term corresponds to additional bandwidth consumption for UE to authenticate the MME through the Internet under TIA-AKA. For simplicity, as long as the UE can receive an IP address and establish the default bearer, it considers that the MME is legitimate.

Table 1. Length of authentication parameters

Parameters	Length (bits)	Parameters	Length (bits)
RES, MAC, Type, TAl, IMSI	64	AMF	16
K (LTE K), RAND, CK, IK	128	SQN, AK	48
KSI_{ASME}	3	K_{ASME}	256
Service request	8	AUTN	128
Physical address	48	AV	608

The simulated network consists of one MME area, dividing into three tracking area (TA). Each TA contains seven eNBs. For TIA-AKA, the authentication processes is done only in its first registration. The following parameters are used: (1) average velocity V for UE; (2) movement direction of UE is uniformly distributed over $[0, 2\pi]$; (3) UEs are uniformly populated with the density within the area, ρ; (4) The radii of eNB area, TA and MME are L_1, L_2 and L_3, respectively. The average number of active mobile crossing the area boundary of length L, is given by $R = \rho V L/\pi$. Note that handover happens when UE is in active mode; Tracking Area Update (TAU) happens when UE is in idle mode; registration happens when MS is switched on or moved from one SN to another.

The simulation covers two scenarios, urban area and suburban area. For urban area, $\rho = 1000$ people/km^2, $V = 40$ km/h, $L_1 = 800$ m. Number of MME is 30. From Fig. 4, we know that $L_3 \approx 4.5 * L_1$. Therefore, the average number of authentication request in the HSS is about 382/s. The total bandwidth consumptions for EPS-AKA and TIA-AIA are $382 * (963 + 608n)$ bps and $382 * (1510 + 393 + 608n)$ bps, respectively. For suburban area, $\rho = 100$ people/km^2, $V = 80$ km/h, $L_1 = 1500$ m. The number of MME is 5. Therefore the average number of authentication request in the HSS is about 24/s. The total bandwidth consumptions for EPS-AKA and TIA-AIA are $24 * (963 + 608 * n)$ bps and $24 * (1510 + 393 + 608 * n)$ bps, respectively.

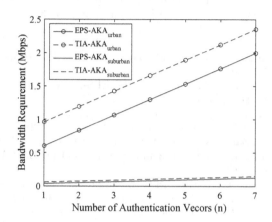

Fig. 6. The bandwidth consumption simulation results of TIA-AKA

From Fig. 6, it can be seen that when more authentication requests occur, the bandwidth consumption of TIA-AKA raises sharply as n grows. Due to the extra message size through MME, the difference between bandwidth consumption of EPS-AKA and that of TIA-AKA increases with the increase of authentication requests. Table 2 compares message sizes of different protocols with $n = 1$.

Table 2. Performance comparison with $n = 1$

Protocol	EPS-AKA	EEPS-AKA	SE-AKA	G-AKA	TIA-AKA
Message bits	1571	1776	2184	1888	2511
Excess percentage	-	13%	39%	20%	60%

6 Conclusion

TIA-AKA is proposed to prevent user identity disclosure and fake eNBs. Efficiency analysis shows that TIA-AKA provides a full protection on the user identity and prevents the DoS attack through the MAC address checkout, at the expense of increased bandwidth consumption and authentication delay. Our future work will be improvement on the efficiency of TIA-AKA with group authentications.

References

1. Abdo, J.B.B., Chaouchi, H., Aoude, M.: Ensured confidentiality authentication and key agreement protocol for EPS. In: RELABIRA 2012 (2012)
2. Abdrabou, M.A., Elbayoumy, A.D.E., El-Wanis, E.A.: LTE authentication protocol (EPS-AKA) weaknesses solution. In: ICICIS 2015, pp. 434–441 (2015)
3. Ahmed, T., Barankanira, D., Antonie, S., Huang, X., Duvocelle, H.: Inter-system mobility in evolved packet system (EPS): connecting non-3GPP accesses. In: ICIN 2010 (2010)
4. Alezabi, K.A., Hashim, F., Hashim, S.J., Ali, B.M.: An efficient authentication and key agreement protocol for 4G (LTE) networks. In: IEEE Region 10 Symposium 2014, pp. 502–507 (2014)
5. Apostol, C.-G., Racuciu, C.: Improving LTE EPS-AKA using the security request vector. In: ECAI 2015 (2015)
6. Cao, J., Ma, M., Li, H., Zhang, Y., Luo, Z.: A survey on security aspects for LTE and LTE-A networks. IEEE Commun. Surv. Tutor. **16**(1), 283–302 (2014)
7. El Idrissi, Y.E.H., Zahid, N., Jedra, M.: Security analysis of 3GPP (LTE) - WLAN interworking and a new local authentication method on EAP-AKA. In: FGCT 2012, pp. 137–142 (2012)
8. Forsberg, D., Huang, L., Tsuyoshi, K., Alanaram, S.: Enhancing security and privacy in 3GPP E-UTRAN radio interface. In: PIMRC 2007 (2007)
9. Gibson, J.D.: Mobile Communication Handbook, 3rd edn. CRC Press, Boca Raton (2013)
10. Gu, L., Gregory, M.A.: A green and secure authentication for the 4th generation mobile network. In: ATNAC 2011, pp. 1–7 (2011)
11. Hamandi, K., Sarji, I., Chehab, A., Elhajj, I.H., Kayssi, A.: Privacy enhanced and computationally efficient HSK-AKA LTE scheme. In: Barolli, L., Xhafa, F., Takizawa, M., Enokido, T., Hsu, H.H. (eds.) WAINA 2013, pp. 929–934. IEEE (2013)
12. ETSI: Digital cellular telecommunication system (phase 2+) (GSM). Universal Mobile Telecommunications System (UMTS), LTE Network Architecture, May 2017
13. ETSI: LTE; general packet radio service (GPRS) enhancements for evolved universal terrestrial radio access network (E-UTRAN) access, October 2017
14. ETSI: Universal mobile telecommunications system (UMTS); LTE; 3G security; specification of the MILENAGE algorithm set: An example algorithm set for the 3GPP authentication and key generation functions f1, f1*, f2, f3, f4, f5 and f5*; document 2: Algorithm specification, April 2017
15. Køien, G.M.: Mutual entity authentication for LTE. In: IWCMC 2011 (2011)
16. Li, X.H., Wang, Y.J.: Security enhanced authentication and key agreement protocol for LTE/SAE network. In: WiCOM 2011 (2011)
17. Mun, H., Han, K., Kim, K.: 3G-WLAN interworking: security analysis and new authentication and key agreement based on EAP-AKA. In: WTS 2009 (2009)
18. Purkhiabani, M., Salahi, A.: Enhanced authentication and key agreement procedure of next generation evolved mobile networks. In: ICCSN 2011, pp. 557–563 (2011)
19. Yu, D., Wen, W.: Non-access-stratum request attack in E-UTRAN. In: ComComAp 2012, pp. 48–53 (2012)

A Smartphone-Assisted Device-to-Device Communication for Post-disaster Recovery

Md. Akbar Hossain[1(✉)] and Sayan Kumar Ray[2]

[1] Department of IT and Software Engineering, Auckland University of Technology, Auckland, New Zealand
akbar.hossain@aut.ac.nz
[2] Faculty of Business and Information Technology, Manukau Institute of Technology, Auckland, New Zealand
sayan.ray@manukau.ac.nz

Abstract. Natural disasters like earthquakes often cause partial or complete breakdown of existing telecommunication infrastructure leaving the helpless people in the affected areas without means of exchanging emergency messages. Under such situations, a temporary ad-hoc system to help in exchanging emergency communication messages and post-disaster recovery can be set up utilising the smartphones of affected victims and the IoT devices of the smarthomes in the affected areas and this paper proposes a method to do that. In the proposed method, smarthome IoT devices are set up to act as relay nodes to communicate emergency messages in absence of a fully functioning telecommunication network. A relay node is chosen based on multiple independent parameters like the residual lifetime of an IoT device and its degree of connectivity. MATLAB-based simulations conducted prove the efficiency of the method.

Keywords: Natural disaster · Energy consumption
Relay · IoT · Smartphone

1 Introduction

During disaster incidents like earthquakes, people get trapped under collapsed buildings or debris and they get seriously injured or even die. Failure to quickly locate and rescue the victims trapped under debris is a problem. Communication of post-disaster emergency messages is highly crucial for fast localization and saving the lives of the victims trapped under rubbles and debris immediately aftermath a natural disaster incident.

Most studies performed so far to provide improved communication abilities during and after disaster situations focused on the deployment of additional wired or wireless connection infrastructure. For example, in [1] a geosynchronous

© ICST Institute for Computer Sciences, Social Informatics and Telecommunications Engineering 2018
P. H. J. Chong et al. (Eds.): SmartGIFT 2018, LNICST 245, pp. 11–20, 2018.
https://doi.org/10.1007/978-3-319-94965-9_2

earth orbit satellite access point (SAP) is deployed in the disaster area to provide connectivity with a cost of time, which is unrealistic in disaster scenarios. A Hybrid Wireless Mesh Network (HWMN) using the free unlicensed spectrum and IEEE 802.11b/a/g off-the-shelf devices is considered in [2]. A combination of the previous concepts proposed in [3] considered a portable transmission tower with two radio interfaces and a Very Small Aperture Terminal transceiver. However, the delay associated with bringing in and deploying a portable communication tower in the disaster area was unrealistic. As an alternative to fixed infrastructure, the authors in [4–6] proposed an on-site network configuration to support disaster recovery based on the concept of wireless Multi-hop communication abstraction. In [6], each smartphone connected to the nearby access point (AP) for communication as well as a virtual AP (VAP) to extend the network. This formed a tree-based multi-hop access network that extended the coverage and provided additional network resources to victims to communicate. Smartphones based disaster recovery methods are also proposed in [7–9] to locate immobilized survivors.

Telecommunication infrastructure during such disaster incidents may partially or completely collapse. In such situations, an ad-hoc communication network may be set using smartphones. A smartphone assisted device-to-device (D2D) victim localisation method (SmartVL) is proposed in [10], where a smartphone self-senses a disaster scenario, self-switches to a pre-set disaster mode and self-connects to nearby available smartphones to create an ad-hoc communication network in order to relay emergency messages containing the tentative location of victims trapped under rubbles. However, SmartVL only considers smartphone-based D2D communication, whereas, currently, there are various other IoT devices that can support communication under such disaster circumstances. So far, only a limited number of efforts [11,12], have focused on the IoT based communication for post-disaster emergency communication and recovery.

In this paper, we propose a smartphone and IoT based D2D ad-hoc networking mechanism to support post-disaster emergency communication and recovery. This research considers that smarthome-based IoT devices can act as relaying devices to relay the emergency messages (data packets) from smartphones belonging to victims in the disaster affected areas to first responders or other rescue people. We consider IoT gateway devices (IoTGD) that can support multiple heterogeneous RATs and normal IoT devices (NIoTD). Such a smartphone and IoT-based multi-hop ad-hoc communication method can be effective post-disaster scenarios with little or no functional telecommunication coverage or internet connectivity. Every IoT device (or node) can choose its immediate relaying device based on the independent parameters like the residual lifetime of the device (depending on the leftover battery energy of the device) and the degree of connectivity of a device enroute the destination. In the remaining paper, the terms 'device' and 'node' are used interchangeably, where a node implies an IoT device.

The rest of the paper is organized as follows: Sect. 2, presents the proposed method, while Sect. 3 explains the selection of an ideal relay node in the method.

The simulation set up and results are discussed in Sect. 4 and Sect. 5 concludes the paper.

2   The Proposed Method

Immediately aftermath a disaster incident, like earthquake, cellular networks in the affected area may get congested owing to an excessive increase in the network traffic volume or can be completely damaged or collapsed leaving helpless people stranded without means of communication. Under such circumstances, smart homes in the disaster affected areas fitted with the different IoT devices (e.g., smart alarms, smart smoke monitors, smart temperature monitors etc.) can be utilised in the emergency communication and recovery as explained here. The different IoT devices in a smart home communicate to each other and relay messages. For the simplicity of this work, we assume that a small scale heterogeneous network in a disaster affected area consists of few structurally symmetric smart homes fitted with the different IoT devices that are fixed and locations of the devices in the homes are known to each other.

This work considers both smartphones and smart homes IoT devices. We assume that smartphones belonging to victims and other people in the affected areas are able to self-monitor the radio environment and detect the occurrence of a natural disaster. Upon detection, a smartphone can self- switch to a pre-defined disaster mode in order to communicate emergency HELP messages to other smartphones in the vicinity. The details of this procedure is discussed in [10]. However, in [10], the authors only considered smartphone to smartphone communication. If a victim or affected person's smartphone is unable to find another smartphone operating in the same mode in the proximity, the communication is dropped and emergency messages are terminated, which certainly is not desirable. To address this issue, in this work we have leveraged the concept of D2D communication and considered the IoT devices in the smart homes to relay the emergency messages in absence of fully functioning cellular networks. We assume that an IoT device can sense an event and communicate messages with other devices (a more powerful IoT device has some processing abilities as well). Also, IoT devices are generally battery powered and its functioning consumes battery energy.

An example of the communication scenario mentioned above is presented in Fig. 1, where a smartphone belonging to user A relays emergency messages to another smartphone belonging to user B through the smart home IoT devices. Here, we assume two different types of IoT devices, namely, an IoT gateway device (IoTGD) and a normal IoT device (NIoTD), which is not a gateway. An IoTGD is a multi-RAT device that can communicate both with smartphones and other IoT devices (i.e, with NIoTDs) and each smarthome can have more than one IoTGDs. A NIoTD on the other hand is just a normal single RAT IoT device that can only communicate to other NIoTDs and IoTGDs but not with a smartphone. Both IotGD and NIoTD are capable of relaying messages. In case of a post-disaster scenario, we further assume that an IoT device can relay emergency messages only if it has adequate amount of leftover energy in the battery to support such actions. This is because, aftermath a disaster, power outage is common and devices have to rely on battery backups for functioning. Therefore, appropriately predicting the leftover energy of a device is important to estimate its tentative life time. An IoT device with leftover energy in the battery below a threshold limit is not considered for the relaying purpose.

In the proposed method, a smartphone which is unable to find another nearby smartphone to pass on the emergency message, can instead communicate the message to an IoTGD in a smarthome (in the disaster affected area) having adequate leftover battery energy. The IoTGD then relays the message to another suitable IoTGD or NIoTD, which has the highest leftover battery energy and the largest degree of connectivity. This process continues until the message is forwarded to another smartphone (located outside the smarthomes) by the final IoTGD. The section below explains the proposed method in detail.

Fig. 1. Considered network scenario

3 Selecting an Ideal IoT Device to Relay Messages

An IoTGD or a NIoTD selects the next IoT device to relay an emergency message based on the following criteria: (a) the lifetime of an IoT device depending on the residual or the leftover energy of the device's battery, and (b) the degree of connectivity of an IoT device.

3.1 Lifetime of an IoT Device

An Iot device can either be in a sleep state or in an active state. While, energy consumption of such a device during sleep state is negligible, it consumes significant energy when operating in an active state. Here we explain the energy consumption of an IoTGD and an IoTD when operating in active state.

Normal IoT Devices: As mentioned above, the cycle of operation of a NIoTD is composed of sleep and active states. The energy consumption while in the sleep state is negligible in comparison to that in the active state and can be written as:

$$E_{sleep} = P_{sleep} \times T_{sleep} \tag{1}$$

On the other hand, a NIoTD performs the following activities when operating in an active state:

- Idle: No event occurred; $E_{idle} = P_i \times T_i$.
- Sensing: A device senses the environment and processes the sensing information; $E_{sense} = P_{se} \times T_{se}$.
- Transmitting: A device transmits the processed sensing information; $E_{transmit} = P_t \times T_t$.

Therefore, the total energy consumption for an NIoTD when in an active state can be calculated as:

$$
\begin{aligned}
E_{NIoTD} &= E_{idle} + E_{sense} + E_{transmit} \\
&= P_i \times T_i + P_{se} \times T_{se} + P_t \times T_t
\end{aligned}
\tag{2}
$$

IoT Gateway Device: An IoTGD is capable of communicating to smartphones and hence is a multi-RAT device supporting heterogeneous connectivity. An IoTGD has multiple transceivers and hence consumes more energy than a NIoTD. The energy consumption of an IoTGD can be written as:

$$
\begin{aligned}
E_{IoTGD-hm} &= E_{idle} + E_{sense} + E_{re-hm} + E_{transmit} \\
&= P_i \times T_i + P_{se} \times T_{se} + P_{re-hm} \times T_{re-hm} + P_t \times T_t
\end{aligned}
\tag{3}
$$

Moreover, an IoTGD may require extra energy to convert or process the emergency messages received from or communicated to a smartphone as the message formats may be different. Thus, the energy consumption of an IoTGD including that for the message conversion is:

$$
\begin{aligned}
E_{IoTGD-ht} &= E_{idle} + E_{sense} + E_{re-ht} + E_{transmit} \\
&= P_i \times T_i + P_{se} \times T_{se} + P_{re-hm} \times T_{re-ht} + P_t \times T_t
\end{aligned}
\tag{4}
$$

It is obvious that $E_{re-ht} > E_{re-hm}$ due to additional processing and protocol conversions of different RAT. Now, for a cycle the total energy consumption is the sum of the energy consumption in sleep state and active state which is as follows:

$$
\begin{aligned}
E_{total} &= E_{sleep} + E_{active} \\
&= \left(E_{sleep} + E_{idle} + E_{sense} \right) \times (1 - R) + R \times \left(E_{transmit} + E_{re-hm} \right. \\
&\quad \left. \times R_{hm} + E_{re-ht} \times (1 - R_{hm}) \right)
\end{aligned}
\tag{5}
$$

Here, the value of R will be 0 for non-relay node and 1 for relay node. Same as if the relay node is homogeneous relay then R_{hm} will be 1 otherwise 0.

Let us consider that E_{thr} is the minimum energy required to maintain the communication with the other IoT devices and relay devices. Hence the lifetime of an IoT device can be defined as a number of cycle periods before the

IoT device reach below the E_{thr}. Hence the lifetime of an IoT device can be written as:

$$K_{life} = max(m) : E_{total} \geq E_{thr} \tag{6}$$

3.2 Neighbour Discovery and Degree of Connectivity

An IoT device needs to discover its neighbour nodes or other available IoT devices to relay messages and also needs to know the degree of connectivity of each of the neighbours. For an IoT device, the degree of connectivity can be defined as the ratio of the number of IoT neighbours it has over a total number of IoT devices [13] in the small network considered. Any IoT device can initiate a neighbour discovery operation by sending a simple HELLO message consisting of the sender's ID, energy level and location coordinate. Upon receiving the HELLO message, the receiver can add the sender as a neighbour and responds back with a HI message with the same. On receiving the reply message, the sender can similarly add the receiver as a neighbour. The neighbour discovery process is presented in Algorithm 1. The degree of connectivity is calculated in line 14 of Algorithm 1.

Algorithm 1. ND Algorithm

Input: (i) Transmission flag, $Flag_{Tx}$;
        (ii) Rescan period, T_{out};
        (iii) No of node, n

Output: (i) The neighbour of node X: $N(X)$;
        (ii) Degree of connectivity, D;

Begin
1: **while** $mod(t, T_{out}) = 0$ **do**
2:     Node X: Send HELLO message
3:     **if** Message received:Node Y **then**
4:         $N(Y) = N(Y) \cup x$
5:         Node Y: Send HI message
6:         **if** Message received:Node X **then**
7:             $N(X) = N(X) \cup Y$
8:         **else**
9:             Eliminate Y from the list
10:        **end if**
11:    **else**
12:        Send HELLO Message
13:    **end if**
14:    $D = \frac{|N(X)|}{n}$
15: **end while**

End

3.3 Relay Node Selection

As mentioned above, in our proposed method, an ideal relay node is chosen based on the lifetime of a node or device (which is dependent on its leftover battery energy) and the degree of connectivity of an IoT device. So, if the neighbour discovery phase, finds two or mode nodes that are eligible to qualify as relay nodes, then ideally the one with the maximum leftover battery energy (i.e., with maximum lifetime) and the highest degree of connectivity is chosen as the relay node. However, the ideal scenario may not always be the case and there could be a trade off between the above two parameters that we may have to consider in order to priorities one node above the other as explained in (Eq. 7) below.

$$T_{factor} = \left(\frac{k}{n}\right)^A + (k_{life})^B \tag{7}$$

Equation 7 ensures that, always the ideal node will be selected as the relay node depending on parameters A and B, which provides the flexibility of choice. If the residual lifetime of a device is the main concern in a disaster scenario, then the node with higher lifetime can be selected, but if the priority is to minimise the message transmission time, then the node with higher degree of connectivity needs to be selected as the relay node.

4   Performance Evaluation

We have used a MATLAB-based simulation system to evaluate the performance of the proposed method and have compared the results with existing relay selection methods explained in [14]. In [14], the authors proposed three relay deployment or selection strategies based on degree of connectivity, lifetime and hybrid. In all cases network operation progress in rounds. Higher the number of relay nodes, lesser is the network lifetime as relay nodes consume more battery power. For our simulation, we have considered three performance parameters, namely, the mean residual energy consumption, relay node survival and average success rate. Table 1 lists the simulation parameters.

Table 1. Simulation parameters

Parameter	Value	Parameter	Value
Number of nodes	50–100	T_{active}	$d \times T_{Cycle}$
Network range	$100\,\mathrm{m} \times 100\,\mathrm{m}$	T_{slp}	$(1-d) \times T_{cycle}$
Data packet size	500 bytes	T_{idle}	$\frac{T_{active}}{2}$ (No event) T_{active} (event)
Control packet size	25 bytes	T_{sense}	$1.1\,\mathrm{s}$
T_{cycle}	$5\,\mathrm{s}$	$T_{transmit}$	$1.4\,\mathrm{s}$
d	$[0,1]$	$E_{threshold}$	$0.5\,\mathrm{mJ}$
$T_{hm} = T_{ht}$	$1\,\mathrm{s}$	$E_{initial}$	$0.5\,\mathrm{J}$

4.1   Mean Residual Energy Consumption

We explain here the performance of our proposed method in context to the mean residual energy consumption parameter, which provides us with an understanding of the residual lifetime of a device. (Figure 2) shows that the mean residual energy in the case of our proposed method is higher than the other protocols. A significant performance difference can be observed with hybrid scheme, which considers both degree of connectivity and node's residual energy. However, the energy consumption in hybrid model is mainly dominant by the distance between relay nodes and base stations. Interestingly, a significant amount of energy is consumed for sensing and data (message) processing in IoTGDs (supporting heterogeneous RATs), which are introduced in our method. Figure 2 shows that mean residual energy of relay nodes in our proposed method is 14% higher than the hybrid, 37% higher than the lifetime and 58% higher than the other connectivity-based relay selection scheme.

4.2   Relay Node Survival

The number of active relay nodes that survive at each round is an important parameter to study in a disaster scenario. Power failure or outages is common aftermath a disaster incident. In such cases, the in-house IoT devices needs to survive on battery backups as long as possible. More the number of active relay nodes, better is the end-to-end delay performance. Figure 3 depicts that proposed method, in context to this parameter, shows the following improvements: 12% in comparison to the hybrid scheme, 27% in comparison to lifetime and 69% in comparison to connectivity schemes. Such improvement is a result of the fact that in our proposed method an IoTGD or a NIoTD only transmits data to the nearest relay node enroute the destination in the multi-hop communication scenario. As higher the distance between two nodes, more is the energy consumption, in our proposed method energy consumption is always less as there are intermediate nodes available to help relay the emergency messages.

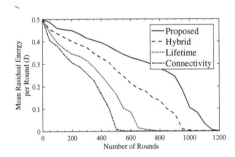   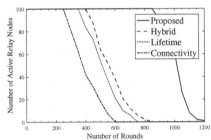

Fig. 2. Mean residual energy with number of rounds

Fig. 3. Relay nodes survival per round

4.3   Average Success Rate

The average success rate (ASR) implies the reliability of the method to successfully transmit packets (messages) to the destination even if intermediate relay nodes fail. ASR is defined as the ratio of the number of transmitted packets from a source to the total number of packets received by the destination from the same source. To study the performance of the proposed method in context to this parameter, the simulation is configured for 1200 rounds. Figure 4 shows that the success rate is increased with the number of relay nodes. In most cases, the proposed scheme has an increased success rate of 10% compared to hybrid, 30% compared to lifetime and 47% compared to connectivity schemes.

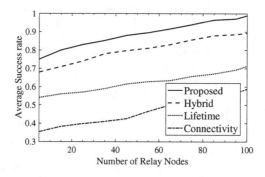

Fig. 4. Average success rate of packet delivery

5   Conclusion

This paper proposes a smartphone and IoT devices-assisted emergency and recovery method in a post-disaster environment, where smartphones can utilise the IoT devices in the smartphones in the disaster affected areas to successfully relay the emergency messages to other smartphones. We considered two different types of IoT devices, namely, the IoTGD and NIoTD, both with relaying capabilities and have proposed methods to select an ideal relaying node based on multiple criteria like, the residual lifetime of an IoT device and the degree of connectivity of each of the devices. Our proposed criterion for relay node selection is appropriate for disaster situations requiring lower energy consumption and end-to-end delay in data transmission. Simulation results have shown better performance of our proposed method in comparison to other such schemes.

References

1. Patricelli, F., Beakley, J.E., Carnevale, A., Tarabochia, M., Von Lubitz, D.K.: Disaster management and mitigation: the telecommunications infrastructure. Disasters **33**(1), 23–37 (2009)
2. Dilmaghani, R.B., Rao, R.R.: A wireless mesh infrastructure deployment with application for emergency scenarios. In: 5th International ISCRAM Conference. Citeseer (2008)
3. Mathur, S.: A rapidly deployable communications network architecture for disaster management. Technical report. Citeseer (2009)

4. Chenji, H., Hassanzadeh, A., Won, M., Li, Y., Zhang, W., Yang, X., Stoleru, R., Zhou, G.: A wireless sensor, adhoc and delay tolerant network system for disaster response (2011)

5. George, S.M., Zhou, W., Chenji, H., Won, M., Lee, Y.O., Pazarloglou, A., Stoleru, R., Barooah, P.: DistressNet: a wireless ad hoc and sensor network architecture for situation management in disaster response. IEEE Commun. Mag. **48**(3), 128–136 (2010)

6. Minh, Q.T., Shibata, Y., Borcea, C., Yamada, S.: On-site configuration of disaster recovery access networks made easy. Ad Hoc Netw. **40**, 46–60 (2016)

7. Nishiyama, H., Ito, M., Kato, N.: Relay-by-smartphone: realizing multihop device-to-device communications. IEEE Commun. Mag. **52**(4), 56–65 (2014)

8. Suzuki, N., Zamora, J.L.F., Kashihara, S., Yamaguchi, S.: SOSCast: location estimation of immobilized persons through SOS message propagation. In: 4th International Conference on Intelligent Networking and Collaborative Systems, INCoS, pp. 428–435. IEEE (2012)

9. Lu, Z., Cao, G., La Porta, T.: Networking smartphones for disaster recovery. In: IEEE International Conference on Pervasive Computing and Communications, PerCom, pp. 1–9. IEEE (2016)

10. Hossain, A., Ray, S.K., Sinha, R.: A smartphone-assisted post-disaster victim localization method. In: IEEE 18th International Conference on High Performance Computing and Communications, pp. 1173–1179. IEEE (2016)

11. Chung, K., Park, R.C.: P2P cloud network services for iot based disaster situations information. Peer-to-Peer Netw. Appl. **9**(3), 566–577 (2016)

12. Sinha, A., Kumar, P., Rana, N.P., Islam, R., Dwivedi, Y.K.: Impact of Internet of Things (IoT) in disaster management: a task-technology fit perspective. Ann. Oper. Res. 1–36 (2017)

13. Dandekar, D.R., Deshmukh, P.: Relay node placement for multi-path connectivity in heterogeneous wireless sensor networks. Proc. Technol. **4**, 732–736 (2012)

14. Xu, K., Hassanein, H., Takahara, G.: Relay node deployment strategies in heterogeneous wireless sensor networks: multiple-hop communication case. In: 2005 Second Annual IEEE Communications Society Conference on Sensor and Ad Hoc Communications and Networks, IEEE SECON 2005, pp. 575–585. IEEE (2005)

Heuristics-Based Detection of Abnormal Energy Consumption

Ankur Sial[1], Amarjeet Singh[1], Aniket Mahanti[2], and Mingwei Gong[3(✉)]

[1] IIIT-Delhi, Delhi, India
[2] University of Auckland, Auckland, New Zealand
[3] Mount Royal University, Calgary, Canada
mgong@mtroyal.ca

Abstract. This paper presents two methods for detecting abnormal electricity consumption by utilizing usage patterns in the vicinity. The methods use contextual and factual information including, energy consumption patterns, nature of supply and category of day to logically group meters and find abnormalities. Using heuristics proposed in the paper, data collected from fifty smart meters deployed inside hostels of IIIT-Delhi were investigated for abnormal electricity consumption. Multiple abnormalities were found and their causes were verified after discussion with campus administrators. Our results show that the proposed heuristics successfully found abnormal energy consumption behavior. Therefore, these methods could be used for real-time abnormality detection. This will result in reducing operating costs by automatically detecting and reporting abnormalities without human intervention.

1 Introduction

Many countries across the globe have started to realize the importance of energy efficiency. Efficient energy usage habits reduce overall expenditure on energy. Energy conservation reduces energy wastage, contributing towards a sustainable development.

Smart meters are the next generation of electricity meters. They help in better energy accounting. Using smart meters, users can make informed decisions for reducing energy usage. Smart meters provide fine grained data that can be used for monitoring, recognition, and profiling of appliances using device signatures.

For efficient energy usage, abnormal behavior should be identified. While developing prediction techniques and benchmarking mechanisms one has to remove anomalous data. Therefore, new methods for abnormal energy consumption detection should be devised to help identify abnormal system behavior. Visualization of energy consumption using smart meters helps in finding abnormal behavior and reduce energy costs. Using smart meters, a better demand response system can be created through pattern analysis.

The main contribution of this paper is to apply two heuristics for detecting abnormal energy consumption detection in a residential campus. Using smart

© ICST Institute for Computer Sciences, Social Informatics and Telecommunications Engineering 2018
P. H. J. Chong et al. (Eds.): SmartGIFT 2018, LNICST 245, pp. 21–31, 2018.
https://doi.org/10.1007/978-3-319-94965-9_3

meters we track energy consumption for hostel residents in the campus. To the best of our knowledge, previous research analyzed energy consumption at a house or a building level. We are the first to detect abnormal energy consumption for a group of meters having same context inside hostels in a residential campus. Our empirical evaluation demonstrates the effectiveness of the proposed heuristics. An abridged preliminary version of the work appeared in [1].

The remainder of this paper is organized as follows. Section 2 discusses related work. Section 3 describes our methodology including data collection and contextual separation of the data set. Section 4 presents two abnormal energy consumption detection heuristics. Section 5 discusses observations and results for each of the heuristics. Section 6 concludes the paper.

2   Related Work

With rapid development of smart grids, researchers have proposed various techniques for abnormal detection in smart grids. A sliding window framework was proposed to integrate historical sensor data and contextual features for detecting anomalous behavior [2]. Saad and Sisworahardjo [7] presented contextual anomaly detection algorithm to detect irregular power consumption using unsupervised learning algorithm and temporal context generated from meter readings. Rossi et al. [6] evaluated an approach for anomaly detection in smart grids derived from data streamed from smart meters. The proposed approach took into account the aspects of collective and contextual anomalies. In [5], the authors introduced load profiling methodology relying on consumption data from smart meters. The non-monitored users' load is then decomposed using artificial neural network trained with the available data.

Chen and Cook [4] transformed time series energy consumption data into a symbol sequence. They used suffix tree to identify occurrence of patterns. Then, the power sequences were clustered to find outliers in the data. They used two month data from apartments to detect energy outliers. We used clustering to get outliers in the data in one of the heuristics. Apart from clustering, we followed a completely different heuristic-based approach to detect abnormal behavior. In [8], the authors used peak energy consumption in daily readings to find abnormal energy consumption in buildings. They used energy consumption history and calculated variation from normal. They used mean and standard deviation to detect abnormal energy consumption. In our approach, we have also used energy consumption history. Apart from that we have used percentage change in consumption and distance from K-nearest neighbor days in heuristic 2 to detect abnormal behavior.

Bellala et al. analyzed data from 39 m in three buildings in a commercial campus [3]. They combined Support Vector Machines (SVM) with Hidden Markov Model (HMM) to propose a semi-supervised approach. They used occupancy models to reduce load on lighting on one floor. Our work is different as we used unlabeled data and unsupervised approach to detect abnormal consumption behavior.

3 Methodology

IIIT-Delhi campus is spread over 25 acres in Okhla, New Delhi, India. After phase 1 of construction, 30,000 m^2 of space was covered by buildings. Schneider Electric EM64XX series meters are deployed across campus and collect over 5 million data points every day. Over 200 of these smart meters are deployed inside 13 buildings in IIIT-Delhi campus. This includes student hostels, lecture halls, academic offices, faculty housing, library building (including student labs), sport arena, and centralized facilities. For our analysis in this paper, energy consumption data from 50 m of girls hostel and boys hostel were used.

3.1 Data Collection

Raspberry Pi controllers are connected to groups of meters over a common RS485 serial communication bus. Using smart controllers, the data is pulled from meters and transported over campus LAN. We utilized the Simple Measurement and Actuation Profile (sMAP) for data storage, transmission, and communication with devices.

Every 30 s, over 10 electrical parameters (including energy, voltage, power factor) are collected by each smart meter. This is in contrast to others, where data is collected at much larger granularity (30 min or above). Data collected at such high resolution allows detailed energy accounting. The data is stored into archive which is queried to access desired data. The dataset used contains energy data from January, 2014 to April, 2015 (16 months). All the days were labelled as weekday#, weekend#, holiday# in chronological order and meters were labelled as power# and backup# where # is the unique identifier (number).

Although, granularity as low as 30 s is useful for finding appliance usage signatures, very low granularity gives unnecessary level of details for energy consumption profiles. Therefore, we initially used aggregated data with hourly granularity to find abnormal energy consumption. Later, detailed causes for abnormal consumption were found using fine grained data.

3.2 Contextual Separation of Data

Energy consumption is dependent on the context. Separating and grouping data with similar context is helpful for enhanced data analysis. Therefore, we grouped the collected data on three parameters: hour of the day, type of the day, and type of supply.

In our case, the energy data is collected from student hostels of a residential campus. For regular students, the campus is functional eight months in a year. Rest of the four months includes winter and summer break of one and three months respectively. During breaks, many hostel wings are not occupied. Therefore, vacation and working days have different consumption patterns.

In IIIT-Delhi most of the classes are scheduled on weekdays. Weekdays and weekends have different consumption patterns as well. This is because hostel residents whose parents live nearby visit their homes during weekends. Thus, working days were further divided into weekdays and weekends. The three types of days are summarized as follows: Working days (Working weekdays and working weekends of the semester), Vacation days (Between semester holidays and mid-semester break), and Weekends (non-working).

To summarize, the data is divided into groups based on the following different contexts:

1. Day-wise context (3): Working days, weekends, and long holidays.
2. Hour of the day context (3): 0000–0600, 0700–1600 and 1700–2300 h.
3. Supply type context (2): Power and light backup.

In total, we formed 18 different groups ($3 \times 3 \times 2$ for each combination of day, hour and supply type). The expected intra-group daily energy consumption pattern is same. Therefore, abnormal energy consumption was analyzed by comparing different meters within the group.

4   Abnormal Consumption Detection Heuristics

We describe our heuristics to detect abnormal or unexpected energy consumption for a group of meters having same context. Using these heuristics, campus administrators can focus only on the anomalous meters and inspect unexpected behavior.

4.1   Heuristic 1: Abnormal Consumption Detection Using Percentage Change in Consumption

For a group of meters having same context, the energy consumption is similar. The actual energy consumption values depend on the behavior and usage pattern of the occupants. In IIIT-Delhi, most of the hostel rooms are occupied by same occupants for two consecutive semesters. Therefore, for a given meter, the change in behavior is dependent on external conditions including weather.

When we compare different meters having same external conditions, the rate of change in consumption is expected to be similar. The data used in this paper is collected from meters deployed at wing level inside two buildings. Every floor of the building has approximately three to five meters. Also, these buildings are located next to each other. In this environment, all meters are expected to exhibit similar behavior.

In other words, the rate of increase and decrease in energy consumption has to be similar for a given day. If most of the meters show decrease in their energy consumption, steep increase in consumption for the rest of the meters is an abnormal behavior.

This heuristic uses the percentage change of energy consumption to calculate abnormal energy consumption score. The input to the heuristic is a quantized set of energy consumption values. For a given day, the input to the heuristic is consumption data of current day and previous day of all meters. This data is chosen from one of the formed groups having same context. Using this heuristic, abnormal consumption score is calculated using the multiple steps. This data set is divided into subsets of a configurable width. For example, hours 0700–1600 can be divided into continuous subsets of length 5 namely 0700–1100, 0800–1200, 0900–1300, 1000–1400, 1100–1500, 1200–1600. Anomalous data for shorter durations of time is helpful to detect smaller segments of anomalies. This is because, while calculating percentage change in energy consumption, longer duration tends to normalize anomalous behavior. In case the abnormal behavior exists for longer durations, the effective score for longer durations will be magnified while combining the scores for a given set. This is because combined scores of subsets are used to find the score of a set. Also, shorter duration anomaly scores are helpful while narrowing down on abnormal behaviors.

For the given day and its previous day, total energy consumption is calculated. Using these values, the percentage change in energy consumption ($P[i]$) is calculated. To differentiate, this value is kept negative for decrease in total energy consumption and positive for increase in total energy consumption. From the values obtained in previous step, two sets of meters, namely increasing meters and decreasing meters, are formed. Meters with increase or decrease in energy consumption are called increasing meters or decreasing meters respectively.

Now, each of the meters in the two sets is multiplied with the ratio of meters in the opposite set to obtain a score $S[i]$. For example, if there are four meters in decreasing group and 1 m in increasing group. Then the percentage decrease for every meter in decreasing group will be multiplied with 0.2 (1/5) and the percentage increase for every meter in the increasing group will be multiplied by 0.8 (4/5). In this step, meters in minority will be multiplied with ratio of meters in majority and vice versa. The higher ratio of majority depicts higher abnormal behavior by the meters in the minority. The lower ratio of minority depicts lower abnormal behavior by the meters in the majority. Majority meters are those meters whose ratio among increasing and decreasing meters is higher. Similarly, minority meters are those meters whose ratio among increasing and decreasing meters is lower.

Finally, the score, $S[i]$ is multiplied with average percentage change of the majority meters and the average percentage change of the minority meters. This is useful in comparing the anomaly score of current day with the anomaly score of other days. While comparing anomaly score of the current day all the values will be multiplied with the same factor. Therefore, for a given day this will not affect relative anomaly scores. Higher average of majority meters depicts higher chances of anomalous behavior for minority meters. Depending on the use case, one could also replace multiplication of scores with average percentage change of the majority and minority meters with their difference. As the sets were divided into subsets, the abnormal energy consumption score for a set could be calculated using the scores of subsets. To find the score for a given set, all its subset scores were averaged.

4.2 Heuristic 2: Abnormal Consumption Detection Using the Distance from K-Nearest Neighbor (KNN) Days

When all the energy consumption days are clustered together, outliers represent anomalous behavior. This is true when we assume the abnormal energy consuming days are significantly less in number compared to normal energy consuming days. In that case, the distance between two normal energy consuming days will be significantly lower than the distance between an abnormal energy consuming day and a normal day. Therefore, if we find the distance from K-nearest neighbouring days, the distance of anomalous days will be significantly higher than normal days. If there is a possibility of more than K-outliers to be close to each other, the distance from the mean of the data could be factored to calculate anomaly scores.

The input to the heuristic is consumption data. This data is chosen from one of the groups having similar context as the meter whose data is under consideration. A subset of data from a group is used to determine K-nearest neighbour days. Following are the configurations that we used to form subsets and find anomaly scores:

1. Consumption by all meters on all days: This is the basic configuration that can be used to find anomalous behavior. One should note that we are using the subset of data from a group out of the eighteen groups we formed. Therefore, by all meters on all days we mean the entire group. This method works well as data used have same context for the consumption data in a group. Also, one can further improve the results by selecting a subset of the data in a group to further narrow down the context. These cases are discussed in the remaining configurations.
2. Consumption by current meter on all days: This configuration can be used when some of the meters have very different consumption from the set. Different behavior from the set is not expected and should be investigated. If one observes such behavior, one could use this configuration to effectively find abnormal behavior. For example, if we consider the lifts in the hostels, these cannot be grouped with regular meters as the use case for the energy consumption in lifts is different than consumption in hostel rooms. Therefore, we can use this configuration to detect abnormal behavior in meters with unexpected or irregular consumption.
3. Consumption by all meters on current day: This is similar to consumption by current meter on all days. Instead of meters, the days having very different consumption from the rest of the days falls in this category. One can find anomalous energy consumption on some special days when the expected consumption is not same as other days in the group using this configuration. This also takes care of finding abnormal behavior when the data of only one day is given. This configuration works on the principle that per occupant consumption of anomalous meter will be farthest from rest of the meters for a given day.
4. Consumption by K-nearest neighbouring meters on all days: This is a special case in which only K-nearest neighbouring meters are used to find K-nearest

neighbour days. The meters in a group have same context for the consumption data. We can further narrow down the context by selecting K-nearest neighbouring meters for every meter. This can ensure that the neighbours used to detect anomalies are closer than other members in the group. This, in turn, ensures improvement in abnormal consumption scores.

5. Consumption by K-nearest neighbouring meters on current day: This case narrows down the context further for the previous case. This case is combination of case 3 and 4. One may choose this when conditions mentioned in both the cases 3 and 4 is true i.e. one needs a narrowed context for a special day.

One can directly use the sum of distances as anomaly scores. Also, one can normalize the scores to compare anomaly scores for different days or heuristics.

Table 1. Hourly consumption data (in Watt hour) for the groups.

Group#	Supply type	Day	Hours	Average usage	Max usage	Min usage
1	Backup	Holiday	Night hours	29.2754	322.6667	0.0035
2	Power	Holiday	Night hours	30.998	338.8333	0.0052
3	Backup	Holiday	Morning hours	29.9433	312.25	0.0035
4	Power	Holiday	Morning hours	29.4218	306.3333	0.0104
5	Backup	Holiday	Day hours	20.2426	317.625	0.0052
6	Power	Holiday	Day hours	24.3739	375.1667	0.0104
7	Backup	Weekday	Night hours	40.9852	415.4167	0.0026
8	Power	Weekday	Night hours	44.264	634.6667	0.0104
9	Backup	Weekday	Morning hours	41.8783	405.7917	0.0035
10	Power	Weekday	Morning hours	40.3534	312.2396	0.0052
11	Backup	Weekday	Day hours	28.843	357.9167	0.0026
12	Power	Weekday	Day hours	32.5202	350.5	0.0026
13	Backup	Weekend	Night hours	37.1992	387.75	0.0026
14	Power	Weekend	Night hours	39.2877	362.8333	0.0026
15	Backup	Weekend	Morning hours	38.808	369.1667	0.0069
16	Power	Weekend	Morning hours	37.4918	301.1667	0.0313
17	Backup	Weekend	Day hours	27.5047	335.9167	0.0026
18	Power	Weekend	Day hours	30.3037	358.8333	0.0052

5   Empirical Evaluation

To detect anomalies, energy data from January, 2014 to April, 2015 (16 months) is used for the analysis process. For the analysis, the energy consumption data from 50 m of girls hostel and boys hostel was used. For boys hostel, total covered area (on ground) and covered area (on floors) is 1116.19 m^2 and 6798.57 m^2 respectively. For girls hostel, total covered area (on ground) and covered area (on floors) is 838.99 m^2 and 3562.28 m^2 respectively. Girls and boys hostel have five and seven floors including ground floor. The hostel data is sampled with granularity of 30 s.

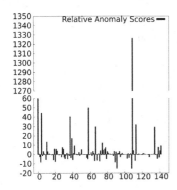

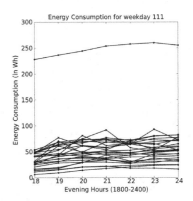

Fig. 1. Anomaly scores for heuristic 1 (light backup)

Fig. 2. High consumption in girls hostel ground floor wing BC

There are two power lines in the hostel namely power and light backup. Every floor has three wings namely A, B, and C. One energy meter collects the energy consumption data of 1/1.5/2 wings of a floor. For example, these could be (i) wing A or (ii) wing A with half of B or (iii) wing A and wing B. The hostels were fully occupied during semesters whereas only some of the floors are occupied during vacations. Therefore, the meters were contextually separated as discussed in Sect. 3.2. The expected behavior of meters within a group was same. This is because the meters are placed inside one building and therefore, the data is collected from exactly the same environment. Also, one should note that the factors like temperature, rainfall, humidity etc. do not play any role as we are comparing relative change in energy consumption for floors in same environment.

5.1   Observations and Results

We discuss observations and results describing effective usage for the heuristics described in the paper. The details about contextually separated groups is described in Table 1. The average and maximum hourly consumption for these groups varies between 20 to 45 and 301 to 635 W respectively. As one can notice that the energy consumption for some meters is twice when compared to others. Therefore, the formation of different contextual groups will help find anomalies effectively.

We next discuss the abnormal behavior highlighted by the heuristics mentioned in the paper.

Heuristic 1: Detection Using Percentage Change

The energy consumption data was directly fed into the heuristic to calculate anomaly scores. Figure 1 presents the anomaly scores for light backup. We can clearly identify several anomalous energy consumption trends (for example the weekday 111) using the heuristic. Upon further investigation, we found out that one of the meters located in girls hostel (ground floor wing BC) was displaying anomalous behavior as shown in Fig. 2. Consumption per occupant for this meter was found to be much higher compared to rest of the meters. This meter provides electricity to 6 occupants, 2 guest rooms, and a common room. Guest rooms are mostly unoccupied. The issue was discussed with the administrators and one of the three power supply phases was found using 4.9 A current (five times greater than the expected behavior which was shown by other two phases).

Such observations can be made real-time by campus administrators and misuse of energy can be pinpointed. Potential misuse of energy consumption includes use of personal appliances such as heater and refrigerators that are not permitted inside the hostel. This approach will also help in identifying sudden change in electricity consumption.

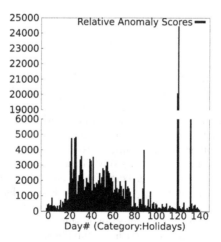

Fig. 3. Morning hours on holidays

Heuristic 2: Detection Using K-Nearest Neighbour Days

For heuristic 2, K = 10 was used to calculate the score. The anomalous days were far away from the given data set and thus, these were detected anomalous by most of the configurations. The campus administrators can select multiple configurations and view the results to find abnormal consumption. Although most of the anomalies are listed by all the configurations, the result is configuration dependent. Also, the order of the anomalies and their priority depends on the configuration used. Therefore, we suggest combining output of all applicable configurations for finding anomalies.

Figure 3 represents anomaly scores for heuristic 2 using configuration 3. Abnormal consumption on day 120 is shown in all three zones: morning, day and night (only morning hours are shown here). Using real time monitoring of morning data, abnormal consumption in day and evening zones can be avoided.

6   Conclusions

We presented two heuristics to identify abnormal energy consumption using contextual grouping of smart meters. Groups were formed by collecting meters with same context. Similar expected behavior were used to compare energy consumption within the neighbourhood of the meters. The heuristics use percentage change in consumption and distance from K-nearest neighbour days to detect abnormal behavior. Data collected from fifty smart meters of hostels of IIIT-Delhi campus were used to analyze these heuristics. Multiple abnormal behaviors were found in the dataset. Our results show that the proposed heuristics successfully found abnormal energy consumption behavior. The abnormal behaviors were verified with facilities administrator of the campus. Our heuristics could be integrated into real-time energy monitoring systems to detect abnormal energy consumption.

References

1. Sial, A., Jain, A., Singh, A., Mahanti, A.: Profiling energy consumption in a residential campus. In: Proceedings of the CoNEXT Student Workshop, pp. 15–17 (2014)
2. Araya, D., Grolinger, K., ElYamany, H., Capretz, M., Bitsuamlak, G.: Collective contextual anomaly detection framework for smart buildings. In: Proceedings of the International Joint Conference on Neural Networks (2016)
3. Bellala, G., Marwah, M., Arlitt, M., Lyon, G., Bash, C.: Following the electrons: methods for power management in commercial buildings. In: Proceedings of the 18th ACM SIGKDD International Conference on Knowledge Discovery and Data Mining, pp. 994–1002 (2012)
4. Chen, C., Cook, D.: Energy outlier detection in smart environments. In: Proceedings of the 7th AAAI Conference on Artificial Intelligence and Smarter Living: The Conquest of Complexity (2011)
5. Ponocko, J., Milanovic, J.: Application of data analytics for advanced demand profiling of residential load using smart meter data. In: Proceedings of the 12th IEEE PowerTech Conference (2017)
6. Rossi, B., Chren, S., Buhnova, B., Pitner, T.: Anomaly detection in smart grid data: an experience report. In: Proceedings of the IEEE International Conference on Systems, Man, and Cybernetics (2016)

7. Saad, A., Sisworahardjo, N.: Data analytics-based anomaly detection in smart distribution network. In: Proceedings of the International Conference on High Voltage Engineering and Power System (2017)
8. Seem, J.: Using intelligent data analysis to detect abnormal energy consumption in buildings. Energy Build. **39**(1), 52–58 (2007)

Real-Time CPU Scheduling Approach for Mobile Edge Computing System

Xiaoyi Yu, Ke Wang$^{(\boxtimes)}$, Wenliang Lin, and Zhongliang Deng

Beijing University of Posts and Telecommunications, Beijing, China
wangke@bupt.edu.cn

Abstract. Mobile edge computing (MEC) system has outstanding advantages of providing smart city applications with relatively low latency and immediately response. How to guarantee the QoS of the services in MEC system is consequently becoming a hot issue. This work focuses on solving the problem by real-time CPU scheduling. The proposed scheduling algorithm considers different services arrival profiles, computation time consumption and deadline requirements simultaneously. Specifically, the combination and optimization of support vector machine (SVM) and earliest deadline first (EDF) algorithm is designed, which could automatically classify services type and efficiently allocate the computation time in real-time manner. By deploying the traffic trace from the real world, the proposed scheduling algorithm could reduce 45% latency and improve the reliability of transmission, comparing with popular fixed-priority CPU scheduling algorithm.

Keywords: SVM · EDF · Low latency
High reliability of transmission

1 Introduction

Mobile Edge Computing (MEC) System, a concept proposed by ETSI in 2014, provides IT and cloud computing capabilities within the radio access networks in close proximity to mobile subscribers [1]. It is widely considered to be the key technology to realize future smart city applications, such as wireless smart grid, Internet of things, etc.

MEC reduce the latency by offloading computation-intensive tasks to the edge cloud. However, the limited computational resource in the edge clouds may result in the Quality of Service (QoS) degradation [2]. Multi-class network traffic classification helps identify the application utilizing network resources, and facilitate the instrumentation of QoS for different applications. Early traffic classification systems rely on transport layer port number to classify flows. However, with the wide use of dynamic ports, the less effectiveness makes the technique based on port number unreliable. Signature matching technique was proposed by Moore [3]. It derives signature patterns from various network traffic flows and classifies the traffic flow through these matching signature patterns. Although

© ICST Institute for Computer Sciences, Social Informatics and Telecommunications Engineering 2018
P. H. J. Chong et al. (Eds.): SmartGIFT 2018, LNICST 245, pp. 32–42, 2018.
https://doi.org/10.1007/978-3-319-94965-9_4

it's classification accuracy is high, the continuous updating of signature patterns and its inability of handling encrypted packets limit the application [4]. Machine learning methods classify traffic flows according to the flow's statistical characteristics (e.g. packet size, flow duration, etc.). In [5], the authors use 12 features for two data sets, the UNB ISCX network traffic data set and their internal data set, to classify by k-NN classification algorithm. Tsinghua university in [6] classify 7 classes of internet applications with 9 feature parameters, and all of them can be obtained from the packet header. These methods provide a guideline to classify the network traffic, but lots of features also increases processing time, leading to serious latency.

In a real-time system, system's performance and throughput are highly affected by CPU scheduling. With the development of smart phones and wearable devices, the problem of finding robust or flexible solutions for scheduling problems is very importance for applications. Out of some important scheduling, Round Robin algorithm is much efficient, which assume that all servers have the same processing performance. An efficient dynamic Round Robin algorithm for CPU scheduling in [7]. However,this scheduling algorithm depends on choosen time quantum and the relationship between time quantum size and process running time. [8] provide a algorithm, which is based on the existing EDF dynamic scheduling algorithm. The algorithm improves the real-time response of EDF to a certain extent by using the comparison of priority and the time slice borrowing strategy, which is disadvantageous to the low utilization rate of idle time slice of EDF algorithm. In fact, the CPU scheduling algorithm for real-time tasks with deadline has been extensively studied in real-time systems. Dynamic-priority-based EDF algorithm is known to be theoretically optimal for scheduling sporadic real-time tasks [9].

To the best of our knowledge, there is no real-time scheduling algorithm for the multi-class network traffic with SVM classification algorithm in MEC system has been proposed. And in existing studies, most of them are just using SVM for basic classification. There are few optimizations for its parameters. In this paper, we will put forward a real-time CPU Scheduling approach for Mobile Edge Computing System, which combine SVM classification algorithm with parameters optimization and EDF Scheduling algorithm for Mobile Edge Computing. 4 classes of network traffic with 2 feature parameters is used to classify. The simulation results show that combining the SVM algorithm and EDF algorithm can reduce computing latency about 45% and improve the reliability of transmission throughput compering with FP scheduling.

2   System Model

2.1   C-RAN Framework Model

C - RAN network evolve from the traditional distributed base station, as a new type of broadband wireless Internet access technology. We choose C - RAN network as an example to introduce our experiment.

Consider a general C-RAN system which consists of three main components: remote radio heads (RRHs), baseband processing units (BBUs), optical transport network. As shown in Fig. 1, there are N RRHs serving N cells in the transmission system and there are M BBUs in the BBU pool.

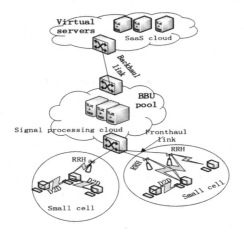

Fig. 1. c-ran framework.

In the process of the BBU pool, PRB increase with the MCS. Then we will build an accurate model to describe the contribution of each underlying BBU functions to the total processing time and how they scale with the increase of PRB and MCS. In this experiment, we consider the three main BBU function: IFFT/FFT, modulation and coding. Generally, there are two important elements in the BBU pool in the course of processing: basic processing and dynamic processing load. The basic processing includes IFFT and FFT for each PRB and the platform-specific processing relative to the reference GPP platform. The dynamic processing load includes user processing, namely coding and modulation, which is the distribution of the PRBs and the linear function of the MCS. On this basis, a model for calculating the total BBU processing time of different PRB, MCS and platforms is proposed, and the following formula is presented as

$$T_{subframe}(z, r, w)[us] = \underbrace{c[z] + p[w]}_{base\,proces\,\sin g} + \underbrace{u_r[z]}_{RMSE} + \underbrace{u_s(z, r)}_{dynamic\,proces\,\sin g} \quad . \quad (1)$$

where the triple (z; r; w) represents PRB, MCS, and platform. The c[z] and p[w] are the base offsets for the cell and platform processing, $u_r[z]$ is the reminder of user processing, and $u_s(z, r)$ is the specific user processing that depends on the allocated PRB and MCS. The $u_s(z, r)$ is linearly fitted to $a(z)r + d(z)$, where a, d are the coefficients, and r is the MCS. Table 1 provide the processing model parameters of the Eq. (1).

However, in our experiment, we chose the LXC as our experiment platform. Considering the practicability of the C - RAN network, the number of PRB that each RRH set configuration is determined by system bandwidth and assigned to each user equipment PRB is derived by the channel status. We assume that these two amount of PRBs are static. Under this promise, we set PRB to 25. Note that MCS 9, 16, and 27 corresponds to QPSK, 16QAM, and 64QAM with the highest coding rate [10]. In our experiment, we randomly picked one of the three values to be the MCS value.

Table 1. Processing model parameters in u_s

z	c	p		$u_s(z,r)$		u_r	
		GPP	LXC	a	d	GPP	LXC
25	23.81	0	5.2	4.9	24.4	41.6	57.6
50	41.98	0	5.7	6.3	70	79.2	80

2.2 Traffic Model

The Professional term and symbol in this work are following the traditional definition in real-time systems. we adopt the most commmonly used traffic model for introducing the real-time analysis into C-RAN, i.e. periodic task with constrained deadline. In each RRH, we assume that there are four types of traffic:video, Browse the web, qq, e-mail. In this section, we use an array of P elements (x_p,y_p) and a mapping f(x) $\rightarrow$ y to describe the packet that we capture, we define X = $\{x_1, x_2, \ldots, x_p\}$ as our traffic flow set, flow x_p properties for classification, $x_p = \{x_{pq}|q = 1 \; or \; 2\}$, q is the number of attributes to class, x_{pq} said the p_{th} packet of the q_{th} properties. Y = $\{y_1, y_2, y_3, y_4\}$to classify categories, respectively, qq, browse the web, video, e-mail. We selected P1 packets from P data packets as training set data for classification function of training.

In our system, we choose four type traffic flow to be observed. We selected two attributes of the data packet, time delta from previous captured frame is defined as t_p, another kind is the length of the packet l_p, among them (p = 1, 2,..., P). P is the total number of packets we capture.t_p and l_p is the x_{pq} which we mentioned above.

First of all, we only run a type of task in the computer, and then capture the network port information. The length of the packet and time delta from previous captured frame of four kinds of network traffic task is shown in Fig. 2. From the Fig. 2, we can clearly see the two attributes of the each task have the obvious difference. The average packet length of video is smaller than the rest. The time delta from previous captured frame of qq is the smallest. The e-mail has the longest packet and browseing the web has its own characteristics too. So, we choose these two properties as attributes for our classification.

For a task x_p, we can use AT_p to express the arrival time (a task start preparing processing time). The deadline is defined as DT_p (the task must be

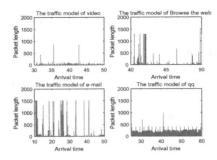

Fig. 2. The traffic model of the four type of task.

completed before the time). For each task x_p, we set a remaining run time RT_p, $RT_p = DT_p$ - t, where t is the current time. For each task x_p, minimum of the RT_p has a higher priority obviously. Obviously, the task x_p which has a higher priority can be processed firstly. x_p can be characterized by three positive integers-worst case execution time $WCET_p$, deadline DT_p, and cycle CT_p, where $WCET_p < DT_p < CT_p$. We set the average of the time delta from previous captured frame as the cycles of four kinds of task. The cycle CT_p of each task is shown in Table 2.

Table 2. Processing model parameters in u_s

Type	qq	Video	e-mail	Browse the web
Cycle/us	0.009203	0.002664	0.003743	0.025081

3   SVM Based Traffic Classification

3.1   SVM Algorithm

We choose SVM as our classification algorithm is because the algorithm can minimize the empirical classification error and maximize set edge classification space. These features reduce the excessive learning the structure of the risk in the limited samples.

Each of our data packets can be expressed as a point on the axis, which can be separated by a line or a plane. We assume that this line or plane is

$$f(x) = w \cdot x + b \tag{2}$$

However, dividing the tasks with only two attributes into four categories can not be done in two-dimensional space, so the kernel function is used in the SVM algorithm. The common kernel functions are linear, polynomial, RBF, etc. We used the RBF kernel function in this experiment. In a multidimensional space, $x \to \varphi(x)$, RBF kernel function can be represented as:

$$K(x_i, x_j) = <\varphi(x_i), \varphi(x_j)> = \exp\left(-g\|x_i - x_j\|^2\right). \tag{3}$$

In order to be able to more accurately classified. We can reduce the problem to:

$$\begin{cases} \min_{w,b} \quad \frac{1}{2}w^2 + C \sum_{p=1}^{p1} \vartheta_p \\ \qquad s.t \\ y_p(w \cdot \varphi(x_p) + b) \geq 1 - \vartheta_p \\ \qquad \vartheta_p \geq 0 \end{cases} \tag{4}$$

The objective function is to maximize the distance between the data points, where ϑ_p is slack variables. The corresponding is that data points x_p allow deviation from the amount of hyperplane. C is penalty coefficient that can limit ϑ_p to infinity. The problem can be solved by Lagrange multiplier. Then Eq. (4) can be converted into Eq. (5) on the dual problem.

$$\begin{cases} max \sum_{p=1}^{P1} \alpha_p - \frac{1}{2} \sum_{I}^{P1} \sum_{J}^{P1} \alpha_i \alpha_j y_i y_j K(x_i, x_j) \\ s.t \quad \sum_{p=0}^{P1} \alpha_p y_p = 0 \quad 0 \leq \alpha \leq C \end{cases} \tag{5}$$

where α_p is the parameter of Lagrange multiplier. By Eq. (5), we can get w and b. The decision function of the final result can be expressed as:

$$f(x) = \omega \cdot x + b = \left[\sum_{p=1}^{P1} \alpha_p y_p K\right] + b \tag{6}$$

In the way, The value of f(x) is represented as one of the type of qq, browse the web, video and e-mail.

3.2   SVM Parameter Optimization

To use SVM, we need to set the parameters. From the above know, we select the RBF kernel function. For RBF kernel functions, we need two functions in general: C and g. For a given problem, we don't know the optimal number of C and g in advance, so we have to search parameter to find the optimal (C, g).

In this experiment, we use the method of grid searching to find the optimal parameters. Because the parallelism of the grid search is very high, and each parameter is independent of each other. The variation range of the penalty coefficient C is $[2^{c\,min}, 2^{c\,max}]$, where we can look for the best C. The default value is $c\,min = -8, c\,max = 8$. Similarly, the variation range of g is $[2^{g\,min}, 2^{g\,max}]$ and the default value is $g\,min = -8, g\,max = 8$. C and g is the horizontal and vertical axes of the grid, cstep and gstep are the step sizes of C and g, which are optimized by the grid parameters. The default values of step size are 1. In this case, The value of C is $\left[2^{c\min}, 2^{c\min} + 1, ...2^{c\max}\right]$ and the value of g is $\left[2^{g\min}, 2^{g\min} + 1, ...2^{g\max}\right]$.

The grid search is to try every possible parameters and use each (C, g) to classify. And then find the best (C, g), which possess the highest precision of cross validation.

Parameter C controls the largest hyperplane and minimizes the data point deviation. We set C to 32768 and g to 8 by cross-validation in our experiment and the accuracy of cross-validation was 79.9729%.

Algorithm 1. Parameter Optimization Algorithm

1: $c = g = 2^{-8}, m = 0$
2: **while** $C < 2^8$ **do**
3:     $C = 2^{-8} + m, m = m + 1, n = 0$
4:     **while** $g < 2^8$ **do**
5:         $g = 2^{-8} + n, n = n + 1$
6:         Use the current g and C for classification. Calculate and record classification accuracy, C and g.
7:     **end while**
8: **end while**
9: To sort the classification accuracy.
10: **return** C and g with the highest classification accuracy.

---

4   Traffic Schedule EDF

In this section we will introduce a preemptive EDF scheduling algorithm.

In numerous real-time scheduling algorithm, the scheduling algorithm that based on priority is one of the most important type of scheduling algorithm in real-time scheduling method. According to the different priority assignment strategy, the scheduling algorithm can be divided into static priority scheduling and dynamic priority scheduling. In general, the dynamic priority scheduling algorithm is better than static priority scheduling algorithm. EDF algorithm is a typical representative of the dynamic priority scheduling algorithm. So we choose EDF algorithm as our scheduling algorithm.

Preemptive EDF scheduling algorithm always performs the first real-time task of the deadline. It is a dynamic priority scheduling algorithm, which is based on the following assumptions:

(1) There is no unpreemptible part of any task, and the cost of preemption can be ignored;

(2) Only the processor requests make sense, memory, I/O, and other resources requests can be ignored;

(3) All tasks are irrelevant; There is no constraint of order;

Based on the assumption of above (1)–(3), the necessary and sufficient condition of the EDF scheduling algorithm for a given periodic task set scheduling is:

$$U = \sum_{p=1}^{P} \frac{T_{subframe}}{CT_p} \leq 1 \tag{7}$$

Thus, the biggest advantage of the preemptive EDF scheduling algorithm is that, for any given set of tasks, as long as the processor utilization is not more than one hundred percent, it can guarantee its scheduling.

As described above, the task x_p is defined by the tuple $\{WCET_p, DT_p, CT_p\}$, Therefore, we need to define how to calculate these parameters in the following. CT_p is shown in Table 2.

$$\begin{cases} WCET_p = T_{subframe}(z, r, w)[us] \\ DT_p = AT_p + T_{subframe}(z, r, w)[us] \end{cases} \tag{8}$$

Algorithm 2. EDF scheduling Algorithm

Input:

CT_p, t(t is the current time), $T_{subframe}$

1: When a data packet arrives, read the arrival time AT_p in the packet
2: Calculate $DT_p = AT_p + T_{subframe}(z, r, w)[us]$
3: Calculate $U = \sum_{p=1}^{P} \frac{T_{subframe}}{CT_p} \leq 1$
4: **while** $U \leq 1$ **do**
5:    **if** $t = AT_p$ **then**
6:        put the data packet into the pending sequence
7: when no new data packet arrives
8: According to DT_p sort from small to large.
9: Deal with the packet with the smallest deadline, the rest packet wait next scheduling
10: **return** The running task

---

5   Simulation

In this part, we study the influence of SVM algorithm and EDF algorithm on the real-time CPU Scheduling approach for Mobile Edge Computing System by simulation.

In the process of classification, We capture 10G data packets by using wireshark through server which network port rate is 5M. The packets contain the four types of task that we will classify. Using cross validation for parameters optimization can improve the classification accuracy of the data. The classification accuracy of the four types of task that we choose is shown in the following Table 3. These tasks is classified by SVM algorithm which use default parameters and parameters optimization respectively.

Table 3. The influence of parametric optimization on classification accuracy

Type	qq	Video	e-mail	Browse the web
The number of packet	196710	849613	22771	63778
Classification accuracy of parameter optimization	0.88726	0.85094	0.84923	0.831525
Classification accuracy of default parameters	0.803594	0.85	0.83536	0.827731

From the Table 3, we can see clearly that using parameter optimization can improve the accuracy of classification.

For scheduling algorithm, in contrast, we select a fixed priority scheduling algorithm, which the scheduling priority of our task is set by the people. In this experiment, we set video as the highest priority, the second is Browse the web, E-mail is the third priority, priority of qq is the last.

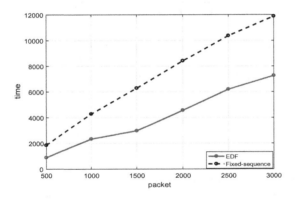

Fig. 3. The influence of different scheduling algorithms on processing time.

In Fig. 3, when we adopt the parameter optimization of SVM classification algorithm and transfer the same number of the packet, we can see that EDF algorithm is better than the fixed priority scheduling algorithm greatly to reduce the time. At the same time, the effectiveness of the EDF algorithm can cut down the processing time and delay, which meet the requirements of transmission.

In Fig. 4, comparing the two kinds of scheduling algorithm of packet loss rate can see clearly that EDF algorithm of packet loss rate is much lower than the fixed priority scheduling algorithm of packet loss rate, which ensure the reliability of transmission.

The simulation results show that the combination of SVM algorithm and EDF algorithm can effectively improve the efficiency of transmission system, satisfy the high reliability and low delay requirement of the requirements of 5G and IOT.

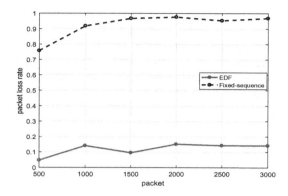

Fig. 4. The influence of different scheduling algorithms on packet loss rate.

6   Conclusions

In this paper, we proposed a real-time CPU Scheduling Approach for Mobile Edge Computing System. It attempts to reduce the latency of the transmission system and the packet loss rate by combining SVM and EDF, and provides a new angle to the scheduling algorithm. The simulation results have illustrated the efficiency of the algorithm.

References

1. Mao, Y., et al.: A survey on mobile edge computing: the communication perspective. IEEE Commun. Surv. Tutor. **19**, 2322–2358 (2017)
2. Zhao, T., et al.: Tasks scheduling and resource allocation in heterogeneous cloud for delay-bounded mobile edge computing. In: 2017 IEEE International Conference on Communications (ICC). IEEE (2017)
3. Jing, N., et al.: An efficient SVM-based method for multi-class network traffic classification. In: 2011 IEEE 30th International Performance Computing and Communications Conference (IPCCC). IEEE (2011)
4. Hao, S., et al.: Improved SVM method for internet traffic classification based on feature weight learning. In: 2015 International Conference on Control, Automation and Information Sciences (ICCAIS). IEEE (2015)
5. Yamansavascilar, B., et al.: Application identification via network traffic classification. In: 2017 International Conference on Computing, Networking and Communications (ICNC). IEEE (2017)
6. Li, Z., Yuan, R., Guan, X.: Accurate classification of the internet traffic based on the SVM method. In: IEEE International Conference on Communications 2007, ICC 2007. IEEE (2007)
7. Farooq, M.U., Shakoor, A., Siddique, A.B.: An Efficient dynamic round robin algorithm for CPU scheduling. In: International Conference on Communication, Computing and Digital Systems (C-CODE). IEEE (2017)
8. Yue, M., Yue-Qi, Z., Zhen-Yu, Y.: Research on real-time scheduling method of RTAI-linux based on edf algorithm. In: 2017 10th International Conference on Intelligent Computation Technology and Automation (ICICTA). IEEE (2017)

9. Pathan, R.M.: Design of an efficient ready queue for earliest-deadline-first (EDF) scheduler. In: Proceedings of the 2016 Conference on Design, Automation and Test in Europe. EDA Consortium (2016)
10. Nikaein, N.: Processing radio access network functions in the cloud: critical issues and modeling. In: Proceedings of the 6th International Workshop on Mobile Cloud Computing and Services. ACM (2015)

Optimal Placement and Sizing of DG and Shunt Capacitor for Power Loss Minimization in an Islanded Distribution System

Mingu Kang and Ramon Zamora[✉]

Electrical and Electronic Engineering Department,
Auckland University of Technology, Auckland, New Zealand
kangiggs89@naver.com, ramon.zamora@aut.ac.nz

Abstract. Active power loss of transmission lines in a distribution system has been a frequent concern for a great number of researchers. Various approaches have been proposed to minimize active power loss. In this research, two approaches, optimal placement and sizing of distributed generation (DG) and shunt capacitor, are used to reduce the active power loss in a distribution system. Particle Swarm Optimization (PSO) and Newton-Raphson method are integrated to find the optimal location and size of DG and shunt capacitor while maintaining operation constraints. The proposed technique is tested on a radial distribution system based on Vava'u island's distribution system in Tonga. The algorithm is implemented in MATLAB and the results are verified with DIgSILENT PowerFactory. Three case scenarios of the optimal placement and sizing, namely: DG only, shunt capacitor only, and DG & shunt capacitor, were tested. The technique successfully found the optimal location and size of DG and shunt capacitor in three case scenarios and further found the most optimal solution among these three cases.

Keywords: Distributed generation · Shunt capacitor
Particle Swarm Optimization (PSO) · Optimal location · Optimal size
Active power loss

1 Introduction

The current power system is in transition from conventional power system to smart grid. The power system is expected to be more environmentally friendly, secure, reliable, resilient, efficient, and sustainable [1]. The levels of transition and achieved expectation are various among systems; some systems are still at very beginning of transition, while others are already very advanced. Regardless of the current system position in the transition range, the smart grid concept will help the system to improve.

This paper deals with a conventional islanded distribution system, based on Vava'u Island distribution system. The system is expected to be improved in terms of minimizing power loss and maintaining all bus voltages above 0.9 p.u. The improvement is

© ICST Institute for Computer Sciences, Social Informatics and Telecommunications Engineering 2018
P. H. J. Chong et al. (Eds.): SmartGIFT 2018, LNICST 245, pp. 43–52, 2018.
https://doi.org/10.1007/978-3-319-94965-9_5

related to two of the defined functions of a smart grid: optimizing assets and operating efficiently [2]. The method to achieve this improvement will be discussed below.

The average electrical power loss is around 8% of output power in transmission and distribution system. The percentage of power loss approximately below 10% which occurs in some countries can be ignored, whereas in some developing countries such as Togo, Libya and Haiti, the power loss is much higher. To deal with the significant power loss in some developing countries, a great number of researchers have proposed different approaches for power loss minimization [3]. In this research, two effective methods will be used to reduce power loss in a distribution system. The first one is the installation of DG and the other one is shunt capacitor as a compensator.

Placing DG has the advantage of reducing the amount of power requested from the transmission system and/or central power plant by supplying local loads. Similarly, placing a shunt capacitor at the load can compensate reactive power loss by supplying reactive current to the load. However, inappropriate placement and sizing of DG and shunt capacitor may increase power loss in a distribution power system. Hence, an optimization method is required to find an optimal solution.

Several methodologies based on analytical and meta-heuristic approaches can be used to find the optimal solution. An analytical calculation by using loss sensitivity factor (LSF) and priority list is implemented in [4]. Zero-point analysis in radial or open loop systems as well as 2/3 rule for power loss are introduced in [5]. Another analytical method for optimal placement of DG based on bus admittance matrix, generation information and load in a radial distribution system is presented in [6]. Genetic Algorithm (GA), Taboo Search (TS), Ant Colony Optimization (ACO), Artificial Bee Colony (ABC) and Harmony Search (HS) have been listed in [7] as heuristic methods for optimal placement of DG or shunt capacitor.

Due to several merits in various applications, Particle Swarm Optimization (PSO) is used in this paper to find optimal location and size of DG and shunt capacitor. PSO requires only a fitness function instead of complex mathematical functions such as gradient or matrix inversion. Due to the simple objective function, the computational time will be short with high quality solutions unlike time consuming algorithms such as TS and GA. Since PSO uses basic mathematical and logic operations as well as has fewer parameters, it is characterized as simple to program and easy to implement. In addition, while some of other algorithms require good initial conditions, it is not necessary to have adequate initial condition for PSO to converge because of population-based characteristics. PSO technique is also so flexible that it can be combined with other optimization techniques [8].

The PSO algorithm will be combined with Newton-Raphson power flow analysis to make sure that the optimal solution satisfies all operation constraints. Three different scenarios, namely: DG only, shunt capacitor only, and DG & shunt capacitor, are simulated in MATLAB and the results from the three scenarios are verified in DIg-SILENT PowerFactory.

2 Proposed Methodology

The methodology for optimal placement and sizing of DG and shunt capacitor is divided into two sections, load flow analysis and optimization. Newton-Raphson method is used for load flow analysis and Particle Swarm Optimization (PSO) algorithm is used to find the optimal solution. Newton-Raphson power flow method is a very common power flow and discussed in detail in power system analysis textbooks, such as [9, 10]. This paper will not discuss about the power flow method, but will just implement it in combination with PSO as shown in Fig. 2.

Particle Swarm Optimization (PSO) algorithm is a nonlinear algorithm motivated by social behavior of flocking birds or schooling fish. PSO is composed of the number of particles (population), the position, and the velocity of particles in the search space [11]. The algorithm begins with the initialization of the population. The next step is based on two processes: computing the particle velocity and updating the particle position. These two processes will give each particle the direction to move toward the final solution. There are several parameters such as w, c_1 and c_2 which should be considered to facilitate the convergence and prevent the explosion of the swarm. Mathematically, the two vectors represent the position and velocity of particle i in N-dimensional search space as shown in Eqs. (1) and (2). The modified velocity can be represented by Eq. (3), while the improved position of particle i is expressed by (4). The PSO concept is shown in Fig. 1.

$$X_i = [x_{i,1}, x_{i,2}, x_{i,3}, x_{i,4} \cdots x_{i,n}], \tag{1}$$

$$V_i = [v_{i,1}, v_{i,2}, v_{i,3}, v_{i,4} \cdots v_{i,n}], \tag{2}$$

$$v_i^{k+1} = w v_i^k + c_1 r_1 (Pbest_i^k - x_i^k) + c_2 r_2 (Gbest_i^k - x_i^k), \tag{3}$$

$$x_i^{k+1} = x_i^k + v_i^{k+1}, \tag{4}$$

where:

v_i^k is current velocity of particle i at iteration k,
v_i^{k+1} is modified velocity of particle i,
x_i^k is current position of particle i at iteration k,
x_i^{k+1} is modified position of particle i,
$Pbest_i^k$ is personal best of particle i at iteration k,
$Gbest_i^k$ is global best of group,
w is the inertia weight,
c_1 and c_2 are acceleration constants,
r_1 and r_2 are randomly generated number ranged from 0 to 1.

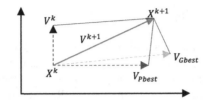

Fig. 1. Concept of PSO search mechanism.

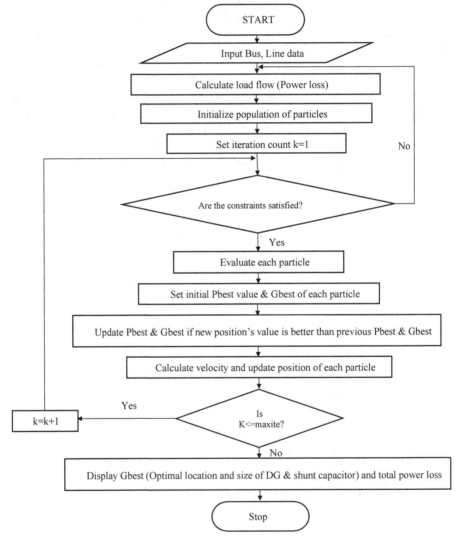

Fig. 2. Proposed PSO flow chart of optimal placement of DG and shunt capacitor.

3 System Design

3.1 Problem Formulation

The simple one-line diagram of radial distribution systems without DG and shunt capacitors is shown in Fig. 3 and with DG and shunt capacitors is shown in Figs. 4 and 5, respectively. The impedance in distribution system represents distribution lines which cause the power loss. The active and reactive power flowing to bus i + 1 is presented in Eqs. (5) and (6).

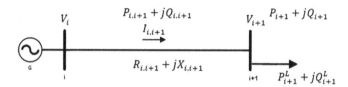

Fig. 3. Simple one-line diagram.

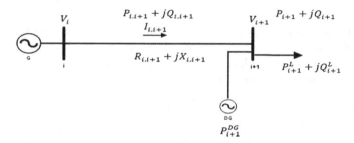

Fig. 4. Simple one-line diagram with DG.

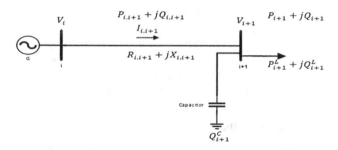

Fig. 5. Simple one-line diagram with capacitor.

$$P_{i+1} = \left[P_{i,i+1} - P_{i+1}^L - \left(R_{i,i+1} \frac{P_{i,i+1}^2 + Q_{i,i+1}^2}{|V_i|^2} \right) \right]. \tag{5}$$

$$Q_{i+1} = \left[Q_{i,i+1} - Q_{i+1}^L - \left(X_{i,i+1} \frac{P_{i,i+1}^2 + Q_{i,i+1}^2}{|V_i|^2} \right) \right]. \tag{6}$$

The objective function of this problem is to minimize active power loss in distribution network as given in Eqs. (7) and (8). All variables in the equations are from the one-line diagram representations in Figs. 3, 4 and 5.

$$F_{ob} = \min P_{TL}, \tag{7}$$

where:

$$P_{TL} = \sum_{i=1}^{NB} R_{i,i+1} \frac{P_{i,i+1}^2 + Q_{i,i+1}^2}{|V_i|^2}. \tag{8}$$

3.2 Systems Constraints

System constraints are necessary measures to ensure system reliability and safety. Voltage limit constraint, line capacity limit and substation capacity constraints should be satisfied to find the best solution of the optimal DG and capacitor placement and sizing. These constraints are given in Eqs. (9)–(12).

$$|V_{min}| \le |V_i| \le |V_{max}|, \tag{9}$$

$$I_i \le I_i^{rated}, \tag{10}$$

$$P_{DG} \le P_S, \tag{11}$$

$$Q_C \le Q_S, \tag{12}$$

where $|V_{max}|$ and $|V_{min}|$ are the maximum and minimum voltage limits at bus i, I_i and I_i^{rated} are the actual current flow and rated maximum current of the conductor, P_{DG} is the active power supplied by DG, Q_C is the reactive power by capacitor, P_S and Q_S are the active and reactive power supplied by substation, respectively.

4 Simulation and Results

The Fig. 6. shows the one-line diagram of the modified 14-bus distribution system based on Vava'u distribution network. It has one central generator. Tables 1 and 2 are the bus data and line data of the system, respectively. The base case analysis is done by Newton-Raphson method to obtain the voltage (p.u), power flow and line loss in the distribution network. The voltage and power loss of base case is shown in Table 3 and Fig. 7.

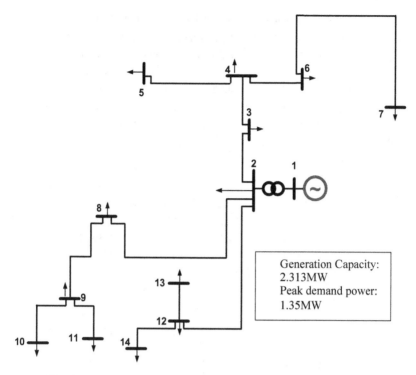

Fig. 6. One-line diagram of the modified 14-bus distribution system.

Table 1. Bus data of the modified 14-bus distribution system

Bus no	Bus code	Voltage mag (p.u)	P(Load) (MW)	Q(Load) (MW)	P(Gen) (MW)	Q(Gen) (Mvar)	Qmin (injected) (Mvar)	Qmin (injected) (Mvar)
1	1	0.0	0.0000	0.0000	0	0	0	0
2	0	0.0	0.5030	0.2436	0	0	0	0
3	0	0.0	0.0451	0.0219	0	0	0	0
4	0	0.0	0.0925	0.0448	0	0	0	0
5	0	0.0	0.1646	0.0797	0	0	0	0
6	0	0.0	0.1306	0.0632	0	0	0	0
7	0	0.0	0.0669	0.0324	0	0	0	0
8	0	0.0	0.0477	0.0231	0	0	0	0
9	0	0.0	0.0547	0.0265	0	0	0	0
10	0	0.0	0.0566	0.0274	0	0	0	0
11	0	0.0	0.0370	0.0179	0	0	0	0
12	0	0.0	0.0873	0.0423	0	0	0	0
13	0	0.0	0.0141	0.0068	0	0	0	0
14	0	0.0	0.0499	0.0242	0	0	0	0

Table 2. Line data of the modified 14-bus distribution system

Line no	Bus no from	Bus no to	Length (km)	R (p.u)	X (p.u)	½ B (p.u)
1	1	2	0.00	0.0000	0.0200	0
2	2	3	1.95	0.0873	0.0336	0
3	2	8	4.92	0.2202	0.0847	0
4	2	12	8.26	0.3698	0.1422	0
5	3	4	1.92	0.0860	0.0331	0
6	4	5	4.75	0.2126	0.0818	0
7	4	6	1.28	0.0573	0.0220	0
8	6	7	8.14	0.3644	0.1402	0
9	8	9	4.63	0.2073	0.0797	0
10	9	10	3.48	0.1558	0.0599	0
11	9	11	3.89	0.1741	0.0670	0
12	12	13	2.55	0.1142	0.0439	0
13	12	14	4.19	0.1876	0.0721	0

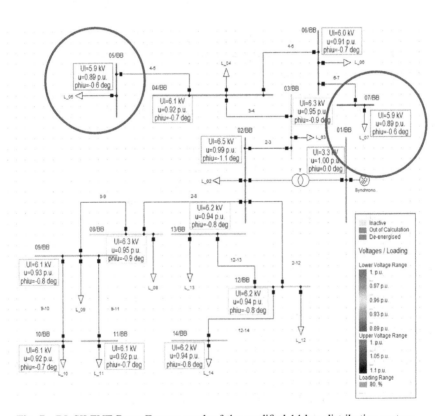

Fig. 7. DIgSILENT PowerFactory result of the modified 14-bus distribution system.

Table 3. Voltage and power flow result of base case

Bus no	Voltage mag (p.u)	Angle (degree)	P(Load) (MW)	Q(Load) (MW)	P(Gen) (MW)	Q(Gen) (Mvar)	Qmin (injected) (Mvar)	Qmax (injected) (Mvar)
1	1.000	0.00	0.0000	0.0000	1.4208	0.7137	0	0
2	0.991	−0.07	0.5030	0.2436	0	0	0	0
3	0.953	−0.89	0.0451	0.0219	0	0	0	0
4	0.918	−0.72	0.0925	0.0448	0	0	0	0
5	0.887	−0.56	0.1646	0.0797	0	0	0	0
6	0.909	−0.67	0.1306	0.0632	0	0	0	0
7	0.887	−0.55	0.0669	0.0324	0	0	0	0
8	0.954	−0.89	0.0477	0.0231	0	0	0	0
9	0.928	−0.76	0.0547	0.0265	0	0	0	0
10	0.920	−0.72	0.0566	0.0274	0	0	0	0
11	0.922	−0.73	0.0370	0.0179	0	0	0	0
12	0.944	−0.84	0.0873	0.0423	0	0	0	0
13	0.943	−0.83	0.0141	0.0068	0	0	0	0
14	0.936	−0.80	0.0499	0.0242	0	0	0	0
Total power			1.3500	0.6538	1.4208	0.7137		
Power loss								0.07082

The maximum DG and capacitor sizes are taken based on the total load demands of 1.35 MW and 0.654 MVAr as shown in Table 3. The variables of population in the PSO algorithm are location and size of DG or shunt capacitor. The randomly initialized variables in each particle are compared and selected for the better position with respect to optimal objective results, the optimal result is the minimum total power loss in the distribution system. The parameters of this PSO algorithm are as follows: maximum number of iteration is 200, population size is 10, inertia weight (w) is 0.9, and acceleration coefficient c_1 and c_2 are 0.5 except for case 3 which has 300 maximum iteration number due to a convergence issue. Table 4 shows the appropriate location and size of DG and shunt capacitor, the total power loss and the voltage improvement at buses 5 and 7, the two buses with voltage below 0.9 p.u in base case, after the installation in different cases. The simulation results show that the proposed algorithm significantly decreased the power loss while improving the weakest bus voltages.

Table 4. Optimal location and size in three cases and total power loss after installation.

Case	Location	DG size (MW)	Capacitor size (MVAr)	Total power loss (MW)	Loss reduction (%)	Bus 5 voltage (p.u)	Bus 7 voltage (p.u)
Base case	-	-	-	0.0708	0	0.887	0.887
Case 1	4	0.5016	-	0.0369	48	0.953	0.953
Case 2	4	-	0.2575	0.0624	12	0.902	0.902
Case 3	4	0.4948	0.2417	0.0298	58	0.967	0.967

5 Conclusion

The optimal location as well as size of DG and shunt capacitor has been applied for power loss minimization in an islanded distribution system. Three case scenarios are presented to compare the active power loss reduction due to the optimal location and size. The Newton-Raphson power flow and PSO are implemented in MATLAB. The results obtained from the Matlab simulations have been verified with DIgSILENT PowerFactory. The results show that the implemented algorithm successfully reduced power loss while maintaining the constraints. The most optimal solution is found for Case 3 with the total power loss of 0.0298 MW. In addition, the lowest bus voltages of 0.887 p.u at bus 5 and 7 in base case are improved to 0.967 p.u in Case 3. Hence, Case 3 is also the best solution for the voltage improvement.

References

1. Zamora, R., Srivastava, A.K.: Controls for microgrids with storage: review, challenges, and research needs. Renew. Sustain. Energ. Rev. **14**, 2009–2018 (2010)
2. Kroposki, B.: Smart grid overview. In: Fall 2012 Federal Utility Partnership Working Group (FUPWG) Meeting (2012). https://www.energy.gov/sites/prod/files/2013/10/f4/fupwg_fall12_kroposki.pdf
3. The World Bank: World Bank Development Indicator (2014). https://data.worldbank.org/indicator/EG.ELC.LOSS.ZS
4. Acharya, N., Mahat, P., Mithulananthan, N.: An analytical approach for DG allocation in primary distribution network. Int. J. Electr. Power Energ. Syst. **28**, 669–678 (2006)
5. Willis, H.L.: Analytical methods and rules of thumb for modeling DG-distribution interaction. In: Power Engineering Society Summer Meeting, pp. 1643–1644 (2000)
6. Caisheng, W., Nehrir, M.H.: Analytical approaches for optimal placement of distributed generation sources in power systems. IEEE Trans. Power Syst. **19**, 2068–2076 (2004)
7. Georgilakis, P.S., Hatziargyriou, N.D.: Optimal distributed generation placement in power distribution networks: models, methods, and future research. IEEE Trans. Power Syst. **28**, 3420–3428 (2013)
8. AlRashidi, M.R., AlHajri, M.F.: Optimal planning of multiple distributed generation sources in distribution networks: a new approach. Energ. Convers. Manag. **52**, 3301–3308 (2011)
9. Saadat, H.: Power System Analysis, 3rd edn. PSA Publishing LLC (2011)
10. Grainger, J.J., Stevenson, W.D., Chang, G.W.: Power System Analysis. McGraw-Hill Education, New York City (2016)
11. Kansal, S., Kumar, V., Tyagi, B.: Optimal placement of different type of DG sources in distribution networks. Int. J. Electr. Power Energ. Syst. **53**, 752–760 (2013)

Applications of Temporal Network Coding in V2X Communications

Xiaoli Xu$^{(\boxtimes)}$, Yumeng Gao, and Yong Liang Guan

School of Electrical and Electronic Engineering,
Nanyang Technological University, Singapore, Singapore
{xu000021i,ygao005}@e.ntu.edu.sg, eylguan@ntu.edu.sg

Abstract. Due to network dynamics and channel fading, multi-hop communication in vehicular networks usually suffers from much higher packet loss rate than the conventional static or single-hop networks. By encoding over packets received at different time slots at the intermediate nodes, temporal network coding (TNC) is a promising technique to avoid erasure accumulation with communication hops. In this paper, we present different strategies of TNC schemes to meet the decoding and delay requirements of different V2X (vehicle to everything) applications. Specifically, for applications with stringent delay requirement, such as live video streaming, we propose to use chunked TNC without precoding. For multi-hop communications with high throughput requirement, we propose to apply carefully designed precoding on top of the TNC to enhance the end-to-end throughput. Different from the conventional TNC code designs, we apply TNC design with overhearing, exploiting the broadcast nature of wireless communication. Specifically, we assume that a vehicle can not only receive the packets from its immediate upstream vehicle, but also overhear some packets from further-upstream vehicles. The number of network coded packets generated at the intermediate nodes is designed by considering the packets received via both the upstream and overheard transmission, which helps to maximize the communication throughput delivered to the destination node.

Keywords: Temporal network coding · V2X communications
Video streaming · Content distribution · Overhearing

1 Introduction

With the emerging technology of vehicular ad-hoc networks (VANETs), vehicles are able to communicate with fixed roadside infrastructure as well as other moving vehicles while traveling on the road - this is referred to as V2X communications. Historically, V2X communications are mainly driven by safety applications such as collision avoidance, emergency warning and crossing in the absence of signaling. Most of the safety applications rely on the periodically broadcasting of basic safety message (BSM), which contains the vehicles' locations, speeds, accelerations and etc. With the increasing popularity of V2X communication system,

© ICST Institute for Computer Sciences, Social Informatics and Telecommunications Engineering 2018
P. H. J. Chong et al. (Eds.): SmartGIFT 2018, LNICST 245, pp. 53–63, 2018.
https://doi.org/10.1007/978-3-319-94965-9_6

the designs of non-safety use-cases to offer drivers enhanced comfort and entertainment have extended the mission of intelligent transportation system (ITS) and attracted significant research interests [1]. The infotainment services, such as the video streaming and location-aware content distribution, usually need to deliver a large amount of information during short-lived communication sessions, which hence impose stringent delay and throughput requirement.

The characteristics of VANETs, such as highly dynamic network topology, error-prone wireless channels and a limited end-to-end bandwidth, impose great challenges on providing high quality safety and non-safety applications. By allowing the information to be encoded at the intermediate nodes, network coding is believed to be a promising technique for tackling the challenges and enhancing the end-to-end throughput of VANETs [2]. In general, network coding can be implemented in two dimensions: spatial and temporal. Spatial network coding (SNC), where the packets coming from different network links are coded together, is useful for alleviating the network bottleneck and it has been applied to enhance the efficiency of BSM exchange via road-side unit (RSU) in VANETs [3]. On the other hand, with temporal network coding (TNC), the packets arriving via the same link at different time slots are coded together. TNC helps to alleviate erasure accumulation across multiple communication hops and hence enhances the end-to-end network throughput. It has been shown that random linear network coding (RLNC) over a sufficiently large number of time slots can achieve the capacity of multi-hop erasure networks [4]. However, as the number of encoded packets increases, the network coding overhead and encoding/decoding complexity grows quickly, which may even overwhelm the benefit of network coding for networks with limited packet length.

In practice, a large file may be divided into multiple blocks and TNC is applied within each block [5], i.e., only the packets within the same block will be coded together, which is referred to as "chunked TNC" in this paper. Then, the resultant network coding overhead and encoding/decoding complexity will be proportional to the coding block size, instead of the original file size. However, when the encoding block size is small, the number of erased packets for each coding block may deviate from its expectation obtained from long-term channel statistics. Therefore, extra redundant packets, over and above those expected from the network capacity, are usually required to ensure that the coding blocks can be decoded successfully with high probability. An effective approach to enhance the end-to-end throughput of TNC without increasing the encoding block size is via applying proper precoding across different coding blocks. For instance, in random annex code [6] and Gamma network code [7], the parity check constraints among different coding blocks are defined based on an erasure correction code, such as low-density parity check (LDPC) code. By designing parity check constraints based on the classical fountain code, the batched sparse (BATS) code introduced in [8] was shown to have negligible reception overhead for any given coding block size. A BATS code consists of an outer code and an inner code. The outer code is an extension of the traditional fountain code to matrix form, and the inner code employs RLNC at the intermediate nodes for packets within the same batch. The BATS outer and inner code can be jointly decoded using efficient belief-propagation algorithm. Note that while the throughput of the BATS network code is enhanced with precoding, the decoding delay may also increase dramatically. Specifically, the "avalanche" effect is observed for decoding the BATS code, i.e., most of packets can only be decoded after receiving sufficient number of packets for decoding the whole file. The application of BATS code for broadcasting to a group of closely-located nodes in wireless ad-hoc network has been considered in [10].

In this paper, we study the application of TNC for V2X communications. Based on the delay and throughput requirements of different vehicular applications, we propose different code designs. For the delay-stringent applications, such as live video streaming, we propose to apply chunked TNC without precoding so that each coding block can be decoded independently after a sufficient number of coded packets is received. To enhance the network throughput, we consider TNC design with overhearing enabled. Specifically, we assume that a vehicle can not only receive the packets from its immediate upstream vehicles, but also overhear the packets from other vehicles, due to the broadcasting nature of wireless communication. We derive explicit expressions for the number of network coded packets to be generated at each vehicle to ensure that its downstream vehicle can decode each coding block with high probability.

On the other hand, for the content distribution applications, e.g., map or promotion video distribution, the file can be used only after it is fully decoded, hence we should minimize the total number of required packet transmissions by applying proper precoding across the information blocks. To this end, we propose to apply the BATS coding scheme. To further enhance the network throughput, the recoding operations in BATS code are designed with the assumption that overhearing is enabled. The design of BATS code for multi-hop wireless communication with unknown channel statistics has been investigated in [9]. However, the recoding operations of BATS code in [9] follows the conventional RLNC without considering the packet overhearing. In [11,12], the number of recoded packets for each batch is optimized according to the number of innovative packets received for that batch in channel with and without memory, respectively. Note that additional computational complexity for rank evaluation of each batch is induced by implementing the recoding design proposed in [11,12]. In contrast, we consider the same number of recoded packets for all the batches at each intermediate node, which is designed to maximize the information delivered with overhearing enabled. To the authors' best knowledge, we are the first to consider BATS code design with overhearing.

Lastly, the application of BATS code for joint V2I and V2V content distribution is discussed, where the content distribution is completed in two phases: the RSU broadcasting phase and the peer-to-peer (P2P) cooperative sharing phase. The rateless nature of BATS outer code allows the RSU to broadcast continuously without repetition, and the BATS inner code enables efficient P2P communication with network coding among vehicles.

2   Multi-hop V2V Communication

As shown in Fig. 1, we consider multi-hop V2V communication where certain file, e.g., live video captured by the camera at V_1 or infotainment content downloaded from the RSU by V_1, needs to be transmitted from the front vehicle V_1 to all other vehicles in the platoon, $V_2, \ldots, V_n$. We assume that the distance between adjacent vehicles V_i and V_{i+1} is constant and denoted by d_i. This is a reasonable approximation in practice when the vehicles are traveling with similar velocity. Further denote by $p_{ji}, j > i$ the packet loss rate experienced by the potential receiver V_j when V_i is transmitting. In general, we have $p_{j_1 i} > p_{j_2 i}$ when $j_1 > j_2$ due to the larger transmitting distance. We assume that the file is partitioned into K packets, each of L bits. We further assume that one common channel that supports transmission rate R_b bits/second is shared by all the vehicles based on time-division method. The vehicles in the platoon are assumed to be time synchronized, and time is slotted so that each packet transmission consumes exactly one time slot, i.e., time slot duration is specified as $T_s = L/R_b$. The packets are transmitted in the form of broadcasting with no feedback.

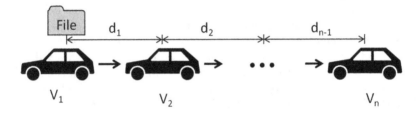

Fig. 1. Multi-hop V2V communication

The end-to-end throughput of this network, ρ, is defined as the ratio between input file size K and the total number of times slots required for delivering the file from V_1 to V_n, which is denoted as T_{total}, i.e.,

$$\rho = \frac{K}{T_{\text{total}}}. \tag{1}$$

The decoding delay, D, is defined as the average number of time slots between a packet being generated at V_1 until it is decoded by the last vehicle V_n. For chunked TNC without precoding, the decoding delay depends on the number of coded packets to be generated for each coding block. On the other hand, due to the "avalanche" decoding effect, the decoding delay for BATS code is equivalent as the total number of time slots required for decoding the complete file, i.e., $D = T_{\text{total}}$.

2.1   Live Video Streaming with Chunked TNC

In this subsection, we consider live video streaming from V_1 to all the subsequent vehicles. To minimize the delay, the video file is divided into small blocks of M

packets, where $M \ll K$, and RLNC is applied within each block. We assume that the network coding coefficients are chosen from a sufficiently large field, and hence the video block can be decoded from any M network coded packets. With chunked TNC, the front vehicle V_1 will send out Y_1 network coded packets when a block of M packets is generated. The number of packets received by V_2 out of the Y_1 coded packets that have been sent out by V_1 is a random variable distributed according to binomial distribution $\mathcal{B}(N, 1 - p_{21})$, which can be approximated by the Gaussian distribution $\mathcal{N}(\mu_{21}, \sigma_{21}^2)$, where $\mu_{21} = Y_1(1 - p_{21})$ and $\sigma_{21}^2 = Y_1(1 - p_{21})p_{21}$. To ensure that V_2 can decode the video block with probability at least $1 - \epsilon$, where $\epsilon \ll 1$, we should have

$$Q\left(\frac{M - \mu_{21}}{\sigma_{21}}\right) \geq 1 - \epsilon, \tag{2}$$

where $Q(x) \triangleq \frac{1}{2\pi} \int_x^\infty e^{-u^2/2} du$ denotes the Gaussian Q-function.

By solving (2), we can find the minimum number of packets to be sent by V_1 as

$$Y_1 = \frac{2M + \alpha^2 p_{21} + \alpha\sqrt{4Mp_{21} + \alpha^2 p_{21}^2}}{2(1 - p_{21})},$$
$$\stackrel{(a)}{\approx} \frac{M + \alpha\sqrt{Mp_{21}}}{1 - p_{21}}, \tag{3}$$

where $\alpha \triangleq |Q^{-1}(1 - \epsilon)|$ and (a) follows from the fact that $\alpha^2 p_{21} \ll M$.

After V_2 successfully decode the block, it will re-generate Y_2 network coded packets from this block, where Y_2 is chosen to ensure that V_3 can decode the block with high probability, together with those overheard packets during the previous Y_1 transmissions. Note that the number of packets overheard by V_3 is a random variable distributed according to $\mathcal{B}(Y_1, 1 - p_{31})$, and the number of coded packets received by V_3 from the dedicated transmission by V_2 is a random variable distributed by $\mathcal{B}(Y_2, 1 - p_{32})$. Therefore, the total number of packets received by V_3 can be approximated by a Gaussian random variable distributed according to $\mathcal{N}(\mu_3 + Y_2(1 - p_{32}), \sigma_3^2 + Y_2(1 - p_{32})p_{32})$, where $\mu_3 = Y_1(1 - p_{31})$ and $\sigma_3^2 = Y_1(1 - p_{31})p_{31}$. To ensure that V_3 can decode with probability $1 - \epsilon$, we can solve for the minimum number of transmissions by V_2 as

$$Y_2 = \frac{M - \mu_3 + \alpha\sqrt{(K - \mu_3)p_{32} + \sigma_3^2}}{1 - p_{32}}. \tag{4}$$

Following the similar analysis, the total number of coded packets overhead from V_i can be approximated by a random variable distributed according to $\mathcal{N}(\mu_i, \sigma_i^2)$, where

$$\mu_i = \sum_{j=1}^{i-2} Y_j(1 - p_{ij}); \quad \sigma_i^2 = \sum_{j=1}^{i-2} Y_j p_{ij}(1 - p_{ij}). \tag{5}$$

Hence, the number of network coded packets generated at all the vehicles can be evaluated in a recursive manner as

$$Y_{i-1} = \frac{M - \mu_i + \alpha\sqrt{(M - \mu_i)p_{i(i-1)} + \sigma_i^2}}{1 - p_{i(i-1)}}. \ i = 2, \ldots, n-1, \tag{6}$$

The network throughput achieved with the proposed chunked TNC is then given by $\rho_{\mathrm{TNC}} = \frac{M}{\sum_{i=1}^{n-1} Y_i}$ and the decoding delay is given by $D_{\mathrm{TNC}} = \sum_{i=1}^{n-1} Y_i$.

2.2 Content Distribution with BATS Code

In this subsection, we consider content distribution from V_1 to the subsequent vehicles. To achieve a good balance between the coding cost and the achievable throughput, we propose to apply BATS coding scheme with the outer code applied at the source V_1 and the inner code applied at the intermediate nodes $V_2, \ldots, V_{n-1}$. Thank to the BATS outer code, we don't have to receive sufficient number of coded packets for every batch to decode the complete file. Therefore, the number of recoded packets should be designed to maximize the useful information delivered, instead of ensuring high decoding probability of each block as in the preceding subsection.

We assume that the content consists of K packets in total. The front vehicle V_1 uses the BATS outer code to generate coded packets in batch, with each batch containing M innovative packets. In the conventional BATS coding scheme, the source node will send out exactly M coded packets and each intermediate node will re-generate M coded packets for each batch from their received packets. To incorporate those packets received via overhearing, we consider a general BATS coding scheme where V_i will send out Y_i coded packets for each batch, for $i = 1, \ldots, n-1$. The number of innovative packets received by V_i for each batch is a random variable denoted by $\mathbf{X}_i$. Denote by η, where $\eta \ll 1$, the coding overhead for BATS code. Then, the original file with K input packets can be decoded from N batches, where $N = \frac{K(1+\eta)}{\mathbb{E}[\mathbf{X}_i]}$ with $\mathbb{E}[\cdot]$ denoting the expectation function. Hence, the end-to-end throughput achieved with the proposed BATS coding scheme is given by $\rho_{\mathrm{BATS}} = \frac{K}{N \sum_{i=1}^{n-1} Y_i}$ and the decoding delay is approximated by $D_{\mathrm{BATS}} = N \sum_{i=1}^{n-1} Y_i$.

Our main task is to maximize the achievable throughput ρ_{BATS} via optimization of $Y_1, \ldots, Y_{n-1}$. However, due to the dependence between those packets received via overhearing and those received via dedicated transmission, it is challenging to characterize the distribution of $\mathbf{X}_i$. For simplicity, we assume that those packets received via overhearing are independent of each other. This assumption is valid when the batch size M is relatively large and the packet loss rates of overhearing links are high. Denote the average number of packets overheard by V_i per batch as O_i, we then have

$$O_i = \sum_{j=1}^{i-2} Y_j(1 - p_{ij}), i = 3, .., n. \tag{7}$$

Further denote by Z_i the average number of packets received by V_i per batch via dedicated transmission from V_{i-1}, and we have $Z_i = Y_{i-1}(1 - p_{i(i-1)})$. Since a batch contains at most M innovative packets, the expected number of innovative packets received by V_i can be estimated as

$$\mathbb{E}[\mathbf{X}_i] \approx \min\{O_i + Z_i, M\}. \tag{8}$$

We assumed that the packets overheard by V_{i+1} form a subspace of $\mathbf{X}_i$ innovative packets received by V_i. Then, the additional number of innovative packets that V_{i+1} can received from the dedicated transmission by V_i is given by $\mathbb{E}[\mathbf{X}_i] - O_{i+1}$. Since the channel between V_i and V_{i+1} has packet loss rate $p_{(i+1)i}$, the expected number of transmissions required for delivering all the additional information is

$$Y_i = \frac{\mathbb{E}[\mathbf{X}_i] - O_{i+1}}{1 - p_{(i+1)i}}, \tag{9}$$

By substituting the initial condition $O_2 = 0$, we can derive the recursive expressions for $Y_i, i = 1, \ldots, n-1$ from (7)–(9) as

$$Y_1 = \frac{M}{1 - p_{21}}, \tag{10}$$

$$Y_i = \frac{M - \sum_{j=1}^{i-1} Y_j(1 - p_{(i+1)j})}{1 - p_{(i+1)i}}, i = 2, \ldots, n-1. \tag{11}$$

The actual throughput and decoding delay of BATS coding scheme depends on the number of batches N required for decoding the file, which further depends on the number of innovative packets received at V_n per batch, i.e., $\mathbb{E}[\mathbf{X}_n]$. It is difficult to obtain explicit expression for $\mathbb{E}[\mathbf{X}_n]$, but it can be easily obtained from simulations.

2.3 Numerical Example

In this subsection, numerical examples are presented to evaluate the performance of the proposed scheme. We assume that there are n vehicles in the network, i.e., $n = 8$ and the neighboring vehicles are separated by equal distance of 50 m. Each vehicle transmits with constant power at 20 dBm. The V2V communication channel has path loss exponent 2.3 and suffer from Nakagami fading with shaping factor 2. A packet can be successfully received if the received SNR is above the receiving threshold of 15 dB. Then, the packet loss rate between V_i and V_j can be obtained as

$$p_{ij} = \begin{cases} 0.0642, & i - j = 1 \\ 0.5987, & i - j = 2 \\ 0.9636, & i - j = 3 \\ 1, & i - j \geq 4 \end{cases} \tag{12}$$

We assume that TNC is applied within a block of 32 packets, i.e., $M = 32$ for both coding schemes proposed in the preceding subsections. The maximum decoding failure probability is set as $\epsilon = 0.01$ in the chunked TNC schemes and the BATS coding overhead is assumed to be $\eta = 0.01$. The achievable throughput for the proposed chunked TNC and BATS coding schemes are compared with the benchmarks where overhearing is not allowed in Fig. 2. As expected, BATS coding schemes usually achieve higher throughput than chunked TNC with the same coding block size, at the cost of larger decoding delay. Furthermore, it is observed that the proposed code designs with overhearing enabled always outperform the benchmark schemes where overhearing is not allowed.

Note that the delay of the chunked TNC is inversely proportional to the achievable throughput as $D_{TNC} = \frac{M}{\rho_{TNC}}$ and hence the proposed chunked TNC with overhearing will have smaller delay as compared to the conventional chunked TNC without overhearing. On the other hand, the delay of the BATS coding schemes is not only related to the coding block size M, but also the original file size K. If a relatively large file is considered, the delay of BATS coding scheme will be much larger than those chunked TNC schemes, regardless whether overhearing is enabled.

3   Joint V2I and V2V Content Distribution

In this section, we briefly discuss the application of BATS coding scheme for content distribution from a single RSU to a group of vehicles passing by it, as shown in Fig. 3. To combat the short connection time and the lossy channel, we propose to further share the contents among the vehicles after they leave the communication range of the RSU. The rateless nature of BATS code allow the RSU to generate unlimited number of coded packets without repetition, and hence greatly simplify the scheduling as compared with the conventional

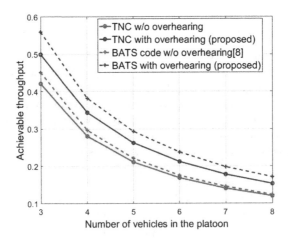

Fig. 2. Comparison of the achievable throughput.

fixed rate coding schemes. Furthermore, the RLNC code within the each batch enhances the efficiency of P2P cooperative sharing. Specifically, the proposed scheme consists of the following two phases:

- *RSU information broadcasting*: The input file is encoded into a set of batches at the RSU and they are sequentially broadcast during the period from the first vehicle entering the RSU communication range until the last vehicle leaving this range. Note that the total transmission time is much larger than the contacting time of a single vehicle with the RSU. Furthermore, due to spatial diversity, the probability that a packet is received by at least one of the vehicles is much larger than the probability that it is received by one specific vehicle. Hence, the number of innovative packets received by the whole group is usually much larger than the number of packets received by one individual vehicle.

- *P2P cooperative sharing*: If the file size is very large such that one vehicle cannot decode the original file based on its own received packets from the RSU, it will broadcast the request for initiating the P2P cooperative sharing phase. In P2P sharing phase, when a vehicle has the chance to access the channel, it will generate and broadcast a coded packet from all the received packets of a batch. The scheduling on which batch should be selected at a given time instance has been designed to maximized the utility of each transmission, with the details provided in [13]. Those packets received from RSU broadcasting phase and P2P cooperative sharing phase can be jointly decoded using belief-propagation algorithm and the P2P sharing phase ends when all the vehicles successfully decode the original file.

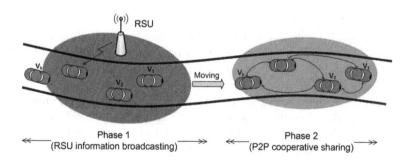

Fig. 3. Joint V2I and V2V content distribution

A numerical example is presented by considering a file with 18 MB distributed by the RSU which is located by the side of the two-lane straight road, to a group of vehicles with relatively close speeds. The packet length is 1500 Bytes. Both the RSU and each vehicle transmits with constant power at 20 dBm. A dual-slop model is adopted to describe the large-scale path loss [13]. The BATS is applied within a batch of 16 packets, i.e., $M = 16$. Since the number of transmissions from the RSU to the vehicles in the first phase depends on the moving speed of the vehicles, we focus on the performance evaluation on the number of transmissions required in the P2P sharing, as shown in Fig. 4. Two benchmark schemes, CodeTorrent [14] and CodeOnBasic [15], adopting TNC, are used for the comparison purposes. Our proposed content distribution scheme with BATS coding requires the lowest number of P2P transmissions to successfully decode the file, significantly outperforming the benchmark schemes.

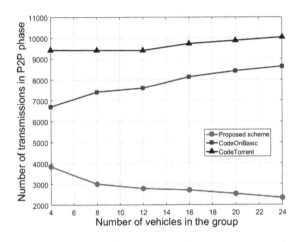

Fig. 4. Comparison of the number of required P2P transmissions.

4   Conclusion

We have discussed the design of TNC, with or without precoding, for different V2X applications to meet their delay and throughput requirements. Multi-hop V2V unicast, as well as joint V2I and V2V content broadcast have been considered. For applications with stringent delay requirement, we considered chunked TNC without precoding. For applications with high throughput requirement, we proposed to use BATS coding scheme. In both cases, the network code designs are augmented with overhearing. The proposed code designs with overhearing are shown to always outperform the benchmark schemes without overhearing.

References

1. Gerla, M., Wu, C., Pau, G., Zhu, X.: Content distribution in VANETs. Veh. Commun. **1**(1), 3–12 (2014)
2. Jamil, F., Javaid, A., Umer, T., Rehmani, M.H.: A comprehensive survey of network coding in vehicular ad-hoc networks. Wirel. Netw. **23**(8), 2395–2414 (2017)
3. Nawaz Ali, G.G. Md., Rahman, A. Md., Chong, P.H.J., Samantha, S.K.: On efficient data dissemination using network coding in multi-RSU vehicular ad hoc networks. In: IEEE Vehicular Technology Conference (2016)

4. Lun, D.S., Medard, M., Effros, M.: On coding for reliable communication over packet networks. Phys. Commun. **1**(1), 3–20 (2008)
5. Chou, P.A., Wu, Y., Jain, K.: Practical network coding. In: Allerton Conference on Communication Control, and Computing, October 2003
6. Li, Y., Soljanin, E., Spasojevc, P.: Effect of the generation size and overlap on throughput and complexity in randomized linear network coding. IEEE Trans. Inf. Theory **57**(2), 1111–1123 (2011)
7. Mahdaviani, K., Yazdani, R., Ardakani, M.: Overhead-optimized gamma network codes. In: International Sympousim Network Coding (NetCod), June 2013
8. Yang, S., Yeung, R.W.: Batched sparse codes. IEEE Trans. Inf. Theory **60**(9), 5322–5346 (2014)
9. Xu, X., Guan, Y.L., Zeng, Y., Chui, C.-C.: Quasi-universal BATS code. IEEE Trans. Veh. Technol. **66**(4), 3497–3501 (2017)
10. Xu, X., Kumar, M.S.G.P., Guan, Y.L., Chong, P.H.J.: Two-phase cooperative broadcasting based on batched network code. IEEE Trans. Commun. **64**(2), 706–714 (2016)
11. Tang, B., Yang, S., Ye, B., Guo, S., Lu, S.: Near-optimal onesided scheduling for coded segmented network coding. IEEE Trans. Comput. **65**(3), 929–939 (2016)
12. Xu, X., Guan, Y.L., Zeng, Y.: Batched network coding with adaptive recoding for multi-hop erasure channels with memory. IEEE Trans. Commun. **63**, 1042–1052 (2017)
13. Gao, Y., Xu, X., Guan, Y.L., Chong, P.H.J.: V2X content distribution using batched network code (2017). https://arxiv.org/abs/1712.00946
14. Lee, U., Park, J.-S., Yeh, J., Pau, G., Gerla, M.: Code torrent: content distribution using network coding in VANET. In: Proceedings of the 1st International Workshop on Decentralized Resource Sharing in Mobile Computing and Networking, pp. 1–5. ACM (2006)
15. Li, M., Yang, Z., Lou, W.: CodeOn: cooperative popular content distribution for vehicular networks using symbol level network coding. IEEE J. Sel. Areas Commun. **29**(1), 223–235 (2011)

Blockchain Based Energy Trading Model for Electric Vehicle Charging Schemes

Chao Liu[1(✉)], Kok Keong Chai[1], Eng Tseng Lau[2], and Yue Chen[1]

[1] Electronic Engineering and Computer Science,
Queen Mary University of London, E1 4NS, London, UK
{c.liu,michael.chai,yue.chen}@qmul.ac.uk
[2] Mechanical, Aerospace and Civil Engineering,
Brunel University London, London, UK
EngTseng.Lau@brunel.ac.uk

Abstract. The electricity market is undergoing a transformation with increasing number of electric vehicles (EV). This paper studies the current charging and discharging schemes for EV and the feasibility of the decentralized transactional energy market. In order to adapt the high volume of EV integration and fully decentralize the grid system, the blockchain technology is introduced to serve as an envision for the next generation grid. Furthermore, we propose a blockchain technology enabled electricity exchange market to enable EVs' autonomy in trading energy with secured distributed energy transactions. The constructed exchange market is price competitive platform where the best bid price is modelled via the jump-diffusion process to enable users to participate in the trading process. We demonstrate that our proposed distributed energy exchange system can perform the peer-to-peer transaction with the real-time electricity price aligning with the EV power demand trend without requiring a third-party intermediary.

Keywords: Electric vehicles · Distributed trading system
Transactional energy · Blockchain technology · Smart grid

1 Introduction

The large-scale integration of renewable energy sources imposes significant challenges to the existing grid systems, mainly because the power generation from renewable energy sources is intermittent and fluctuating. The uncertain power generation and load in the grid lead to unpredictable fluctuation in both the power demand for traditional power generators. Thus, the smart grid is envisaged to be the next generation power grid which combines the stand alone microgrids and large-scale electric power plants [1]. It enables two-way flows of electricity and information to create an automated and distributed advanced energy delivery network that is capable of preserving the stability and resilience of the grid system [2]. However, the fast increasing adoption of EVs brings both challenges

© ICST Institute for Computer Sciences, Social Informatics and Telecommunications Engineering 2018
P. H. J. Chong et al. (Eds.): SmartGIFT 2018, LNICST 245, pp. 64–72, 2018.
https://doi.org/10.1007/978-3-319-94965-9_7

and opportunities to the power grid. On one hand, the massive load demand caused by the integration of EVs into the power grid burdens the power transmission substation and severes the randomness and voltage stability. On the other hand, EVs can be used as mobile energy storages with the bi-directional charging and discharging features to provide ancillary services to the grid system, such as load flattening, peak shaving and frequency fluctuation mitigation [3]. It is, therefore, the vehicle-to-grid (V2G) and grid-to-vehicle (G2V) concepts have introduced that act as the provision of energy and ancillary service from an EV to the grid system [4].

In order to minimize the impact of random behavior from the EV integration, various centralized control scheduling techniques are proposed to provide the ancillary services. In [5], it proposes an optimal scheduling and load curtailment problem for the microgrids to support the islanded operation mode where the parallel computation is used to run the optimisation problem. The proposed scheme ensures the minimal amount of load curtailment while maintaining the reliable operation. Moreover, a stochastic program that incorporates the risk management in [6] is proposed to provide frequency regulation service with the aid of EVs and an aggregator. In [7], an aggregation-based optimisation model for EV charging strategy was proposed with the consideration of stochastic features of the charging procedure in arrival time and state-of-charge (SOC). However, the aforementioned schemes rely on predictions of energy consumption and a day-ahead profile based on historical power consumption and user profiles. In [8], the proposed scheme also lacks of individual decision making process and undermines the autonomy of the individual grid participants. In [9], a distributed trading platform is proposed based on blockchain to support decentralized market approach with facilitate distributed optimization and control. In [10], the authors concluded that a more sophisticated dynamic grid infrastructure can advance the small-scale generators and overall resilience. In [11], a transactional energy is proposed where a sequence of energy transactions for a delivery of a quantity of defined energy product in a specific time interval and location to simplify business for all parties including generators and Distributed System Operator (DSO). Thus, the distributed energy trading platform is based on the clear and frequent communication of offers and transactions among the electricity consumers and operators respectively.

Based on the aforementioned studies, the trading energy in the distributed system is capable of achieving demand response by providing incentivizes EVs to supply and consume electricity of their own self-interest [12]. The advantage of the market-based trading concept is that it reduces the dependency of agents on the DSO or aggregator, as energy supply and demand are matched directly between network peers which results in a more competitive environment. In this regard, this paper proposes a blockchain based transactional energy trading model for the smart grid components, including EVs. This model simulates the trading depth and energy market profile with the best price guide economic concept and allows EVs to charge and discharge autonomy in the grid.

The remainder of this paper is as follow. In Sect. 2, it presents the state-of-the-art in blockchain enabled energy market compositions. In Sect. 3, the concept model is introduced for the distributed energy market with components in smart grids and then further propose the trading model for energy transaction. Then, the simulation results for a local area trading platform with EV market and the concept model evaluation are presented in Sect. 4. Section 5 includes and identifies future research opportunities.

2   Existing Works on Blockchain Enabled Smart Grid

The blockchain is a shared and trusted distributed ledger technology that permits the recording of any digital asset transaction between parties over a decentralized encrypted network which is initially developed as a mechanism to record financial transaction [13]. It confirms transactions in real time and ensures the integrity of transactions through the secured encryption techniques. Henceforth, the blockchain has generated broad interest in other business sectors including the energy trading where all energy traders are the peers in the blockchain network.

The blockchain technology enables a trustless network to eliminate the operation cost of the intermediary participation, which will realise a quicker, safer and cheaper way in the transactional energy market. In [14], a novel mechanism for trading the energy based on the blockchain technology was proposed to adapt the decentralised and competitive environment for the locally produced energy, but the blockchain is solely used as a data storage warehouse to record transactions. In [15], the authors further analyzed the economic evaluation of the market mechanism for local energy trading.

The use of smart contract in blockchain technology is driven by open-source agreements, which also provides the potential to balance supply and demand in the transactional energy market. Furthermore, in [16], the authors provided the insight of the smart contract to allow the automation of multi-step processes to self-execute the distributed and heavy workflows, which is envisaged in the energy industry and the Internet of things.

In summary, the uncontrolled EV charging/discharging may lead to instability of the overall grid system operation. Therefore, it is critical to deploy the effective scheduling algorithm for efficient distributed grid operations on the blockchain based trading platform.

3   Blockchain Based Energy Trading Model for EV Charging Schemes

The transformation to the decentralized transactional energy market can be achieved based on the small-scale energy generators and EVs, in which they may produce, consume, and sell excess electricity capacity like a commodity. It does not require hierarchical system structure, no information exchange, instead, it offers the energy transaction and the agreements on transactions. Hence, all the loads, such as the residents, offices and plants, in the grid are connected to both the retail for end users and wholesale market for large generator offers.

This paper proposes a blockchain based energy trading model that allows prosumers to trade energy in the grid. The proposed model enables the autonomy of prosumers in blockchain power exchange platform, which can inject and draw energy order to the smart grid public blockchain trading platform.

3.1   System Model

In blockchain based energy trading model, we define components as all the power generators and power load components that connected to the retail markets. Each component is capable of publishing and transmitting the charging or discharging order to the smart grid public blockchain trading platform. For EVs, the charging and discharging process can be realized by a programmable charge installation. This is to enable the instant on/off switching of the power transmission as instructed by EVs (assuming the sophisticated design of switches). The energy providers in the public blockchain power exchange platform are the conventional large power plants, distributed micro renewable generators, the storage which compose the electricity provider side and EVs. Besides, the power loads, for example the residential area, hospital, and also EVs are all connected to the public blockchain power exchanging platform.

The information exchange in the blockchain platform is at 30-min intervals. And the components are capable of deciding the price for their produced energy to incentivize users to balance the supply and demand, in the meantime, to reduce the power generation and consumption peaks. The conventional power generators are connected to the wholesale market which trades with large power demand offers, depicted in Fig. 1. Besides providing the wholesale market in the conventional grid system, transactional energy offers a vision for the coordination of retail customers using large numbers of frequent tranching transactions executed automatically by blockchain enabled platform, therefore reducing the centralized features of the next generation grid system [17]. The information exchange is the same for a large generator, distributed energy resource, renewable energy generators such as wind and solar, EV, microgrid, energy trader, broker, exchange, aggregator or system operator. The transactions can be executed between retail and wholesale markets which equalizes the opportunity for all components. Furthermore, the transactions must also account for the transmission and distribution limits and other physical constraints on the grid.

We first define the EV status matrix X as:

$$X_{i,t} = \begin{cases} 1, \text{ if } EV_i \text{ is connected at time } t \\ 0, \text{ otherwise} \end{cases} . \tag{1}$$

The power demand of EV depends on the battery residual (SOC_{ini}) in each EV and the expected SOC (SOC_{exp}) after charging. Hence, it can be formulated as follows:

Fig. 1. Transactional energy market system model with retail and wholesale markets in smart grids where arrow represents the price offers and transactions.

$$P_{EV}(t) = \sum_{i=1}^{I} \left(X_{i,t} \left(SOC_{exp}(i) \pm SOC_{ini}(i) \right) \right). \tag{2}$$

Then, we can define the total residential load as the sum of EV charging/discharging demand and load profile without EV in order to formulate the EV charging problem.

$$P_{total}(t) = P_{load}(t) + P_{EV}(t), t \in T, \tag{3}$$

where P_{load} is the power load generated or consumed by the load within the microgrid network. With the overall utility function for the local area, we can apply optimisation techniques to achieve the objective, such as load flattening, peak shaving and privacy preserving.

3.2   Energy Trading Model

For a electricity exchange order published in the trading platform, the demand is formatted as an input to send to the electricity exchange stand book Std_{in} which is a public order book for all participants in the trading market in the form of a vector which can be denoted as follows:

$$\vec{O_i} = (\gamma, Id_i, \sigma_i, Q_i), \tag{4}$$

where the Id_i is the unique identifier for the charging/discharging initiators where they can be EVs or other components, the σ_i is the unit price that the participant is willing to pay for the electricity order, the Q_i is the electricity demand quantity of this order, and γ is a matrix indicating whether it is a electricity buy or sell order:

$$\gamma = \begin{cases} 1, \text{ buy order} \\ 0, \text{ sell order} \end{cases}. \tag{5}$$

Then for each inserted order, the matched order should be applied to the current stand book (Std_{in}) to generate the matched trades. And all non-error output (each matched trade order) should be directed to the Std_{out}. The trade information format is expressed as follows:

$$\overrightarrow{T_i} = (Id_{sell}, Id_{buy}, \sigma_m, Q_m), \tag{6}$$

where the Id_{sell} and Id_{buy} are the matched electricity buy and sell order identifier respectively, the σ_m is the matched price in pence and the Q_m is the matched quantity for the order. Following the receipt of an order message, and after receiving any matches in the book and outputting any generated trade messages, the solutions should display the current full order book in the above format.

In order to process the transactional energy orders smoothly and ensure the participant benefit, it is crucial to provide the guide price for this demand. We infer from the stock price model based on [18] to construct the best charging guide price S_t with a jump-diffusion process, because the electricity trading market is price competitive which is similar to the stock market. For $S_t < \bar{S}$, where $\bar{S}$ is the highest price for the order, it can be denoted as follows:

$$S_t = S_0 \exp\left((\mu - \frac{\sigma^2}{2})t + \sigma W_t\right), \tag{7}$$

where the percentage drift μ and the percentage volatility σ. Based on the limit scope of this paper, the percentage drift and volatility can be set to constants, and the W_t is a Wiener process. Thus, for a given highest price value S_0, we can obtain the best price S_t by taking derivatives to both sides which is shown as the following equations.

$$dS_t = \mu S_t dt + \sigma S_t dW_t, \text{with } S_0 < \bar{S}, \tag{8}$$

and we can obtain the expectation and variance for S_t, where the expectation can be used as the guide price in the trading process.

$$E(S_t) = S_0 e^{\mu t}. \tag{9}$$

The price function S_t is subject to the highest and lowest price, and fluctuates according to the users' bidding price. It works as the guide price for all participants where all of them are suggested not exceed the guide price. Moreover, they still have the autonomy to decide their trading price between the lowest price and the guide price. It is a closed double auction market with price-time precedence and discrete marketing closing time. Therefore, no central entity is needed to implement the market trading.

4 Simulation Results

To evaluate the performance of the proposed blockchain based energy trading model, a residential area substation transformer with $P_{max} = 250$kVA power capacity is used which serves the size of 100 households. We assume that on average each household would have owned one EV. Moreover, the EV charge connection status is modeled as two parts, where the first time segment is from 06:00 to 18:00 and the second time segment is from 18:30 to 05:30 (+1). In this model, the initial battery residual (SOC_{ini}) for EVs is randomly generated. In order to evaluate the performace of the designed trading market, we generate the electricity buy and sell orders from EV integrated smart grid network. Then the system simulate the exchange process with the order input to calculate the overall price fluctuation with respect to the real-time price.

The price of electricity exchange market is variated according to the guide price order execution where the drift of the best bid price has been assumed to be a constant. To keep the setup tractable for exposition, we assume the simplified scenario: the best bid price exhibits a zero drift $\mu = 0$ prior to the submission of the iceberg order. The original price fluctuation interval is set to be $\sigma_i \in (10, 30)$ subject to the local area, henceforth, the order price σ_t is modified for certain hours during the day to simulate the retail electricity prices σ_{ave} in distribution networks, which are displayed in Fig. 2. As we can see from the figure, the electricity price is higher during 6:00 to 8:00, 11:00 to 13:00 and 17:00 to 19:00, which conforms to the higher power demand P_{EV} for EV charging period as depicted in Fig. 2.

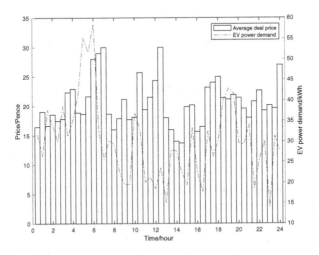

Fig. 2. Average generated trading price in a day.

5   Conclusions

In summary, this paper presents an initial proof-of-concept of energy trading on the blockchain platform. It decentralizes the central controller based smart grid system and increases the autonomy of the grid participants. We provide insight into the economic evaluation of a blockchain-based market design and its technical implementation. In the future work, we will formulate the overall utility function and optimization techniques to achieve the ancillary services.

References

1. Liu, C., Lau, E.T., Chai, K.K., Chen, Y.: A review of wireless power transfer electric vehicles in vehicle-to-grid systems. In: Lau, E.T., et al. (eds.) SmartGift 2017. LNICST, vol. 203, pp. 98–107. Springer, Cham (2017). https://doi.org/10.1007/978-3-319-61813-5_10
2. Moslehi, K., Kumar, R.: A reliability perspective of the smart grid. IEEE Trans. Smart Grid $1(1)$, 57–64 (2010)
3. Liu, C., Chai, K.K., Tseng Lau, E., Wang, Y., Chen, Y.: Optimised electric vehicles charging scheme with uncertain user-behaviours in smart grids. In: 2017 IEEE 28th Annual International Symposium on Personal, Indoor, and Mobile Radio Communications (PIMRC) - Track 4 on "Services, Applications and Business" (IEEE PIMRC 2017 Track 4)
4. Han, S., Han, S., Sezaki, K.: Development of an optimal vehicle-to-grid aggregator for frequency regulation. IEEE Trans. Smart Grid $1(1)$, 65–72 (2010)
5. Nguyen, H., Khodaei, A., Han, Z.: A big data scale algorithm for optimal scheduling of integrated microgrids. IEEE Trans. Smart Grid $\mathbf{PP}(99)$, 1 (2016)
6. Yao, E., Wong, V.W.S., Schober, R.: Risk-averse forward contract for electric vehicle frequency regulation service. In: 2015 IEEE International Conference on Smart Grid Communications (SmartGridComm), pp. 750–755, November 2015
7. Zheng, J., Wang, X., Men, K., Zhu, C., Zhu, S.: Aggregation model-based optimization for electric vehicle charging strategy. IEEE Trans. Smart Grid $4(2)$, 1058–1066 (2013)
8. Cazalet, E.G.: Transactional energy market information exchange (TeMIX), An OASIS Energy Market Information Exchange Technical Committee White Paper (2010). http://www.cazalet.com/images/Transactional_Energy_CW_2010_Cazalet.pdf. Accessed 9 Aug 2012
9. Münsing, E., Mather, J., Moura, S.: Blockchains for decentralized optimization of energy resources in microgrid networks. In: 2017 IEEE Conference on Control Technology and Applications (CCTA), pp. 2164–2171, August 2017
10. Ilic, D., Silva, P.G.D., Karnouskos, S., Griesemer, M.: An energy market for trading electricity in smart grid neighbourhoods. In: 2012 6th IEEE International Conference on Digital Ecosystems and Technologies (DEST), pp. 1–6, June 2012
11. Oh, S.C., D'Arcy, J.B., Arinez, J.F., Biller, S.R., Hildreth, A.J.: Assessment of energy demand response options in smart grid utilizing the stochastic programming approach. In: 2011 IEEE Power and Energy Society General Meeting, pp. 1–5, July 2011
12. Mihaylov, M., Jurado, S., Avellana, N., Moffaert, K.V., de Abril, I.M., Nowé, A.: Nrgcoin: Virtual currency for trading of renewable energy in smart grids. In: 11th International Conference on the European Energy Market (EEM14), pp. 1–6, May 2014

13. Aste, T., Tasca, P., Matteo, T.D.: Blockchain technologies: the foreseeable impact on society and industry. Computer **50**(9), 18–28 (2017)
14. Mihaylov, M., Jurado, S., Van Moffaert, K., Avellana, N., Nowe, A.: Nrg-x-change a novel mechanism for trading of renewable energy in smart grids, pp. 101–106, 01 2014
15. Mengelkamp, E., Notheisen, B., Beer, C., Dauer, D., Weinhardt, C.: A blockchain-based smart grid: towards sustainable local energy markets. Comput. Sci. - Res. Dev. **33**(1), 207–214 (2017). https://doi.org/10.1007/s00450-017-0360-9
16. Christidis, K., Devetsikiotis, M.: Blockchains and smart contracts for the internet of things. IEEE Access **4**, 2292–2303 (2016)
17. Patterson, B.T., Geary, D.E.: Real-time transactional power management in a microgrid mesh network: the enernet. In: 2016 IEEE International Telecommunications Energy Conference (INTELEC), pp. 1–7, October 2016
18. Esser, A., Mönch, B.: The navigation of an iceberg: the optimal use of hidden orders. Financ. Res. Lett. **4**(2), 68–81 (2007). http://www.sciencedirect.com/science/article/pii/S1544612306000742

Energy-Efficiency Maximisation in Random Cognitive Radio Networks

Saifur Rahman Sabuj[1], Md Akbar Hossain[2(✉)], and Edmund Lai[2]

[1] Electrical and Electronic Engineering, Brac University,
Dhaka, Bangladesh
srsabuj@bracu.ac.bd
[2] Department of IT and Software Engineering,
Auckland University of Technology, Auckland, New Zealand
{akbar.hossain,edmund.lai}@aut.ac.nz

Abstract. Energy-efficient cognitive radio network has received considerable attention recently because of improving spectrum and energy efficiency. In light of such observations, we present a model for cognitive radio network based on stochastic geometry theory where transmitters and receivers are distributed according to Poisson point process. In this paper, we focus on the optimization problem where energy efficiency is maximized under the constraint of outage probability for primary network and secondary network. We show that the energy efficiency is increased with the increment of threshold for primary network. However, the energy efficiency is maximum for a certain value of threshold in secondary network.

Keywords: Cognitive radio network · Stochastic geometry
Poisson point process · Energy efficiency

1 Introduction

With the increasing demand for new wireless services and applications, as well as the increasing demand for higher capacity, wireless networks have become highly heterogeneous. In this context, it is essential to have updated information on the radio environment to enhance overall network performance. To support these increasing demands, researchers and engineers have made numerous attempts to introduce new technologies and different network deployments. The concept of cognitive radio (CR) is one of them, which facilitates the flexible usage of the radio spectrum [1]. Cognitive radios are being considered as an alternative to current wireless devices in many applications such as smart grid [2], public safety [3], cellular networks [4], and wireless medical networks [5]. Besides the technology, the topology and network deployment also play a significant role to meet these ever increasing demand. Therefore recent practice of cellular network, the base station (BS) deployment shifted from deterministic model (Wyner, hexagonal

© ICST Institute for Computer Sciences, Social Informatics and Telecommunications Engineering 2018
P. H. J. Chong et al. (Eds.): SmartGIFT 2018, LNICST 245, pp. 73–81, 2018.
https://doi.org/10.1007/978-3-319-94965-9_8

and square lattice model) to a random model [6]. Another motivation of moving towards random deployment is energy efficiency. The authors in [7] reported that, the BS is a cellular networks consume the major portion of total energy consumption and contribute more than 70% of the electric bill. Moreover, higher energy consumption results not only a rise in the CO_2 emission also a significant amount of harmful radiation. A recent study found that the global CO_2 share of the ICT is 5% and which is rising at a fast pace, due the increasing number of network applications, services, and subscribers [8]. Therefore, an energy efficient green wireless communication becomes a prerequisite.

In order to study the system performance such as energy efficiency, outage probability, coverage probability and interference in the CR networks, the stochastic method is commonly used in the literature [9–12]. In [9], the authors uses the Poisson, Binomial, Hard Core point process and Poisson cluster process to evaluate the performance of multi-tier CR networks. However, the authors in [10] shows that the use of Poisson point process (PPP) is impractical for assessing interference and outage probability due to the interaction between the primary and CR users through exclusion region. The authors in [11] investigate the outage probability in two-tier cognitive heterogeneous cellular networks using the theory of homogeneous PPP. Following this line of thought, the authors show that outage probability can be decreased by the proper choice of the spectrum sensing in femto cell. In [12], authors discuss the coverage probability and transmission rate in Rayleigh-lognormal fading.

In this work we have analysed the performance of CR network based on the theory of stochastic geometry. The main contribution of this work is two-fold: (a) We maximize the energy efficiency of primary network under the constraint outage probability of primary receiver. (b) We maximize the energy efficiency of secondary network under the constraint outage probability of secondary receiver. Finally, we present the numerical results of optimal transmission power and density. Also we show the maximum energy efficiency for different threshold. The rest of the paper is organised as follows. In Sect. 2, we present the network model and timeslot structure that has considered in this work. In Sect. 3 we first describe the outage probability for primary receiver (PR) and secondary receiver (SR) followed by energy efficiency for primary network (PN) and secondary network (SN). An optimisation problem is also discussed to maximise the energy efficiency for both PN and SN. In Sect. 4, the numerical results and discussion for a CR network are provided. Finally we conclude our work in Sect. 5.

2   Network Model

Consider a downlink scenario of CR network consisting of secondary users (SUs) and primary users (PUs) is shown in [13,14]. In CR network, two types of network are comprised such as primary network (PN) and secondary network (SN) or cognitive radio network (CRN)[1]. Primary transmitters (PTs), primary receivers

---

[1] cognitive radio and secondary user are used interchangeably.

(PRs) and secondary transmitters (STs) are included in PN and also STs, secondary receivers (SRs) and PTs are included in SN. All active STs and PTs are assumed to be spatially distributed in accordance with homogeneous PPPs Φ_{st} and Φ_{pt} with densities λ_s and λ_p, where λ_s and λ_p are the average numbers of STs and PTs per unit area, respectively. Consequently, the associated SRs and PRs are located in accordance with independent PPPs Φ_{sr} and Φ_{pr} with densities λ_{sr} and λ_{pr}, respectively. We assume a scenario of constant transmit power, and denote by P_s and P_p the transmit power of STs and that of PTs, respectively.

In this work, we have considered time slotted network operation on licensed channel. Predominately, PUs are the legitimate users if the licensed channel and SUs can only use it when it is not used by PUs. We assume that the activity of PUs follows the two state Markov model which are ON and OFF period. The ON period indicates the presence of PUs, hence channel is busy. The OFF periods refers the absence of PUs, thus channel free and can be used by SUs. In order to detect the free slots the SUs, first perform spectrum sensing at the beginning of each frame during the time interval of $[0, T_s]$. This sensing result is used to create a list of available channels that can be used as a potential data channel. An SU will select the best channel from the available channel list based on QoS requirement and transmit data during the time interval of $[T_s, T_s + T_p]$.

3   Analysis of Performance Metric

In this section, the outage probability and energy efficiency have provided based on [13,14] where outage probability and energy efficiency have discussed in details. This paper focuses on the optimization problem. We maximize the energy efficiency under the constraint of outage probability. In the downlink of CR networks, the outage probability depends on the signal to interference plus noise ratio (SINR) of each receiver (i.e., PR or SR).

3.1   Outage Probability

In wireless communication, the outage refers to a significant performance degradation if the SINR is less than a certain threshold. The outage probability can be defined as the probability of SINR being less than the threshold.

Outage Probability of PR. A PR is successfully decoded a packet when it's SINR is above the threshold θ_p. Therefore, the outage at the PR occurs whenever the SINR at the PR declines below a threshold θ_p. According to [13,14], the outage probability for PR can be expressed as

$$\epsilon_p = 1 - \frac{\lambda_p}{\lambda_p - \lambda'_p - p_{im}\lambda_s\beta^2 + 0.5\nu}, \tag{1}$$

where $\nu = \mathbb{L}_p V(\theta_p, 4) + \mathbb{L}_s W_b(\theta_p, 4, \beta)$, $V(\theta_p, 4) = \Gamma(-1/2, \mu_p\theta_p g_p) - \Gamma(-1/2)$, $W_b(\theta_p, 4, \beta) = \Gamma(-1/2, \mu_p\theta_p g_s P_s P_p^{-1}\beta^{-4}) - \Gamma(-1/2)$, $\mathbb{L}_p = \lambda'_p\sqrt{\mu_p g_p \theta_p}$,

$\mathbb{L}_s = \lambda_s p_{im} \sqrt{\mu_p g_s \theta_p P_p^{-1} P_s}$, and $p_{im} = p_{ic,im} p_{st,im}$. P_p is the transmission power of PT, P_s is the transmission power of ST, θ_p is the threshold, λ'_p is the density of PTs located in the area $\mathbb{R}^2 \setminus b(0, r_p)$, $b(x, y)$ represents a sphere of radius y centered at point x, g_p is the interference channel gain between the tagged PR and the interfering PT, g_s is the interference channel gain between the tagged PR and the interfering ST, β is a constant parameter that depends on the cell radius, $p_{ic,im}$ and $p_{st,im}$ are the *probability of unoccupied channel selection* and the *probability of successful transmission* for imperfect detection, respectively. p_{im} is related with N, M, N_s described in [14] where N is the total number of PU channels, M is the number of unoccupied channels and N_s is the sensed channels from the unoccupied channels M, respectively.

Outage Probability of SR. The SUs are unlicensed users in this network and only transmit a signal if the channel is free. The outage probability of a SR can be defined as one minus coverage probability of the SR. The coverage probability for a SR is the probability that the SINR of the tagged SR is greater than the threshold θ_s. According to [13,14], the outage probability for SR can be expressed as

$$\epsilon_s = 1 - \frac{\lambda_s}{\lambda_s - p_{im}\lambda'_s + \nu_1}, \tag{2}$$

where $\nu_1 = p_{ic,pt}\lambda_p\Gamma(1/2)\sqrt{\mu_s\theta_s G_p P_s^{-1} P_p} + 0.5 p_{im}\lambda'_s\sqrt{\mu_s\theta_s G_s}X(\theta_s, 4)$, $X(\theta_s, 4) = \Gamma(-1/2, \mu_s\theta_s G_s) - \Gamma(-1/2)$, and $p_{im} = p_{ic,im}p_{st,im}$. θ_s is the threshold, λ'_s is the density of STs located in the area $\mathbb{R}^2 \setminus b(0, r_s)$, G_s is the interference channel gain between the tagged SR and the interfering ST, G_p is the interference channel gain between the tagged SR and the interfering PT and $p_{ic,pt}$ is the probability of selecting an unoccupied channel from the total number of PU channels.

3.2 Energy Efficiency

In order to improve the energy consumption, it is important to understand the relationship between energy efficiency and area spectral efficiency. In general, the energy efficiency can be defined as the amount of transmitted bits per unit energy consumption and the amount of transmitted bps per unit bandwidth is known as area spectral efficiency. A different definition is presented in [15] where energy efficiency is defined as the ratio of area total throughput to area total power consumption and the area spectral efficiency is defined as the area total throughput per unit bandwidth per unit area. Hence the energy efficiency ratio (η) can be written as the ratio of area spectral efficiency over average network power consumption.

Energy Efficiency of PN. The energy efficiency for the downlink channel is expressed as

$$\eta_{PN} = \frac{(1 - \epsilon_p) \log_2(1 + \theta_p)}{p_{ic,pt} P_p + P_{pc}}$$

$$= \frac{\lambda_p \log_2(1 + \theta_p)}{\left\{\lambda_p - \lambda_p' - p_{im}\lambda_s \beta^2 + 0.5\nu\right\}(p_{ic,pt} P_p + P_{pc})}, \tag{3}$$

where P_{pc} is the constant circuit power consumption in the PT. For simulation purposes, we assume $P_{pc} = 50$ kW [16].

Energy Efficiency of SN. It is considered that the SN consists of ST, SR and PT. The energy efficiency can be expressed as

$$\eta_{SN} = \frac{(1 - \epsilon_s) \log_2(1 + \theta_s)}{\mathbb{P}_{pc}}$$

$$= \frac{\lambda_s \log_2(1 + \theta_s)}{\left\{\lambda_s - p_{im}\lambda_s' + \nu_1\right\}\mathbb{P}_{pc}}, \quad \text{(b/J/Hz)} \tag{4}$$

where $\mathbb{P}_{pc} = N_{trx} P_o T_t + p_{im}\Delta_s P_s T_p + P_{ss}T_s + P_s T_p$ and $T_t = T_s + T_p$. T_s is the sensing period, T_p is the transmission period, P_{ss} is the power consumption during the sensing period, N_{TRX} is the number of transceivers, P_o is the load-independent power consumption at a nonzero output power, which depends on the circuit power consumption of the four types of BS (i.e., macro, micro, pico, and femto). Δ_s is the slope of the load-dependent power consumption, which is the transmission power consumed in the RF transmission circuits for the four types of BS. From the power consumption expression, we observed that P_o and Δ_s are fixed for any given BS. We can verify two aspects (i.e., N_{TRX} and P_s) for use in the design of an energy-efficient network.

3.3 Optimization Problem

In this section, we introduce the optimization problem of PN and SN which are discussed below:

Maximization of Energy Efficiency of PN. In PN, the energy efficiency is maximized where outage probability of PR is constraint. In this optimization problem, we determine the P_p, λ_p, λ_p' that maximize the energy efficiency. The optimization problem can be formulated as

$$\max_{P_p, \lambda_p, \lambda_p'} \quad \eta_{PN}(P_p, \lambda_p, \lambda_p') \tag{5a}$$

$$\text{s.t.} \quad C1 : \epsilon_p(P_p, \lambda_p, \lambda_p') \le \epsilon_{pp} \tag{5b}$$

$$C2 : \lambda_p' \le \lambda_p \tag{5c}$$

$$C3 : P_p \ge 0, \lambda_p \ge 0, \lambda_p' \ge 0. \tag{5d}$$

In (5), constraint C1 ensures that the outage probability ϵ_p does not exceed the particular value of ϵ_{pp}. C2 verifies that λ_p' is less than λ_p. C3 confirms the non-negative value of P_p, λ_p, and λ_p'.

The optimization problem is difficult to solve mathematically. However, we can solve it numerically using the *fmincon* function in MATLAB software [17]. The objective function in (5a) is an increasing function and the feasibility condition is satisfied. Moreover, there are optimal solutions (e.g., P_p, λ_p, and λ_p') that satisfy the constraints (5b)–(5d). The numerical findings are discussed in detail in Sect. 4.

Maximization of Energy Efficiency of SN. In SN, the energy efficiency is maximized where outage probability of SR is constraint. In this optimization problem, we determine the P_s, λ_s, λ_s' that maximize the energy efficiency. The optimization problem can be formulated as

$$\max_{P_s,\lambda_s,\lambda_s'} \quad \eta_{SN}(P_s,\lambda_s,\lambda_s') \tag{6a}$$

$$\text{s.t.} \quad C1 : \epsilon_s(P_s,\lambda_s,\lambda_s') \leq \epsilon_{ss} \tag{6b}$$

$$C2 : \lambda_s' \leq \lambda_s \tag{6c}$$

$$C3 : P_s \geq 0, \lambda_s \geq 0, \lambda_s' \geq 0. \tag{6d}$$

In (6), constraint C1 ensures that the outage probability ϵ_s does not exceed the constant value of ϵ_{ss}. C2 verifies that λ_s' is less than λ_s. C3 confirms the non-negative value of P_s, λ_s, and λ_s'. Following the above optimization, the optimal solution (i.e., P_s, λ_s, and λ_s') is found by the same procedure. The details of this optimization are presented in Sect. 4.

4   Numerical Results and Discussion

In this section, based on the previous analysis of optimization problem, we present numerical results to evaluate the performance of energy efficiency. In the following numerical results, simulation parameters are provided as $N = 25$, $M = 10$, $N_s = 15$, $T_s = 1$ ms, $T_p = 4$ ms, $\Delta_s = 2.8$, $P_o = 84$ W, $P_{ss} = 0.2$ W, $\mu_p = 0.5$, $\mu_s = 0.5$, $\beta = 1$, $\epsilon_{ss} = 0.1$ and $\epsilon_{pp} = 0.1$.

Figure 1 shows the optimal value of P_p in dBm for different value of threshold θ_p considering the value of densities $\lambda_s = 25/\text{km}^2$, $\lambda_s' = 20/\text{km}^2$ and $P_s = 30$ dBm. The optimal P_p is fluctuated from $\theta_p = 5$ dB to $\theta_p = 22$ dB. Then the optimal P_p is stable. As shown in Fig. 2, as θ_p increases, the optimal value of λ_p slightly increases, but λ_p' decreases.

Figure 3 illustrates that the energy efficiency increases with the increment of θ_p at $\theta_p = 27$ dB and then energy efficiency decreases with the increment of θ_p. The maximum energy efficiency is 1.3×10^{-4} at $\theta_s = 27$ dB. Obviously, as θ_p increases, power consumption reduces so energy efficiency is maximized.

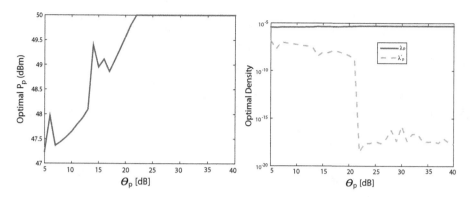

Fig. 1. Optimal P_p for different θ_p in Eqs. (5a)–(5d).

Fig. 2. Optimal density (i.e., λ_p and λ'_p) for different θ_p in Eqs. (5a)–(5d).

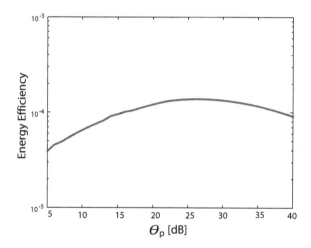

Fig. 3. Energy Efficiency for different θ_p in Eqs. (5a)–(5d).

Now we explore the energy efficiency of SN considering the value of densities $\lambda_p = 15/\text{km}^2$, $\lambda'_p = 5/\text{km}^2$ and $P_p = 50\,\text{dBm}$. A careful observation of Fig. 4 indicates that the optimal value of P_s is dramatically changing. In Fig. 5, similar behaviors can be observed for density (λ_s and λ'_s). The optimal λ_s is constant for different θ_s. But, the optimal value of λ'_s is fluctuated.

Figure 6 shows the relationship between energy efficiency and θ_s. Interestingly, we observe that the energy efficiency is increased at $\theta_s = 7.5\,\text{dB}$ then energy efficiency is decreased with the increment of θ_s. For the optimal setting of P_s, λ_s and λ'_s, the maximum energy efficiency is achieved for different value of θ_s. The maximum energy efficiency is 0.65 at $\theta_s = 7.5\,\text{dB}$.

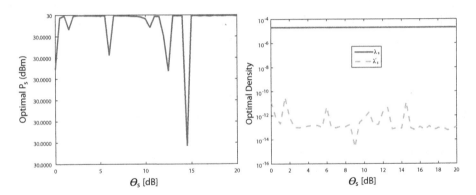

Fig. 4. Optimal P_s for different θ_s in eqs. (6a)–(6d).

Fig. 5. Optimal density (i.e., λ_s and λ'_s) for different θ_s in eqs. (6a)–(6d).

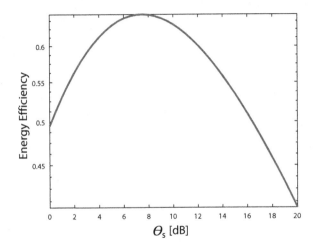

Fig. 6. Energy Efficiency for different θ_s in eqs. (6a)–(6d).

5   Conclusion

In this paper, we have investigated the optimization problem for PN and SN in CR network by using the theory of stochastic geometry. In particular, we maximize the energy efficiency for PN and SN in term of outage probability. Also, we find the optimal transmission power and density of PN and SN for the network that can achieve the maximum energy efficiency. Numerical results demonstrate that the energy efficiency increases in an increase of threshold. For the SN, the energy efficiency is maximum for a particular value of threshold.

References

1. Mitola, J., Maguire, G.Q.: Cognitive radio: making software radios more personal. IEEE Pers. Commun. **6**(4), 13–18 (1999)
2. Ogbodo, E.U., Dorrell, D., Abu-Mahfouz, A.M.: Cognitive radio based sensor network in smart grid: architectures, applications and communication technologies. IEEE Access **5**, 19084–19098 (2017)
3. Gorcin, A., Arslan, H.: Public safety and emergency case communications: opportunities from the aspect of cognitive radio. In: IEEE Symposium on New Frontiers in Dynamic Spectrum Access Networks, pp. 1–10 (2008)
4. Akyildiz, I.F., Lee, W.-Y., Vuran, M.C., Mohanty, S.: A survey on spectrum management in cognitive radio networks. IEEE Commun. Mag. **46**(4), 40–48 (2008)
5. Chávez-Santiago, R., Nolan, K.E., Holland, O., De Nardis, L., Ferro, J.M., Barroca, N., Borges, L.M., Velez, F.J., Goncalves, V., Balasingham, I.: Cognitive radio for medical body area networks using ultra wideband, IEEE Wirel. Commun. **19**(4) (2012)
6. Guo, A., Haenggi, M.: Spatial stochastic models and metrics for the structure of base stations in cellular networks. IEEE Trans. Wireless Commun. **12**(11), 5800–5812 (2013)
7. Han, C., Harrold, T., Armour, S., Krikidis, I., Videv, S., Grant, P.M., Haas, H., Thompson, J.S., Ku, I., Wang, C.-X., Le, T.A., Nakhai, M.R., Zhang, J., Hanzo, L.: Green radio: radio techniques to enable energy-efficient wireless networks. IEEE Commun. Mag. **49**(6), 46–54 (2011)
8. Fehske, A., Fettweis, G., Malmodin, J., Biczok, G.: The global footprint of mobile communications: the ecological and economic perspective. IEEE Commun. Mag. **49**(8), 46–54 (2011)
9. ElSawy, H., Hossain, E., Haenggi, M.: Stochastic geometry for modeling, analysis, and design of multi-tier and cognitive cellular wireless networks: a survey. IEEE Commun. Surv. Tutor. **15**(3), 996–1019 (2013)
10. Lee, C.-H., Haenggi, M.: Interference and outage in poisson cognitive networks. IEEE Trans. Wireless Commun. **11**(4), 1392–1401 (2012)
11. ElSawy, H., Hossain, E.: Two-tier HetNets with cognitive femtocells: downlink performance modeling and analysis in a multichannel environment. IEEE Trans. Mob. Comput. **13**(3), 649–663 (2014)
12. Sabuj, S.R., Hamamura, M.: Random cognitive radio network performance in Rayleigh-lognormal environment. In: 14th IEEE Annual Consumer Communications & Networking Conference (CCNC), pp. 992–997 (2017)
13. Sabuj, S.R., Hamamura, M.: Energy efficiency analysis of cognitive radio network using stochastic geometry. In: IEEE Conference on Standards for Communications and Networking (CSCN), pp. 245–251 (2015)
14. Sabuj, S.R., Hamamura, M.: Outage and energy-efficiency analysis of cognitive radio networks: a stochastic approach to transmit antenna selection, Pervasive and Mob. Comput. **42**, 444–469 (2017)
15. Xin, Y., Wang, D., Li, J., Zhu, H., Wang, J., You, X.: Area spectral efficiency and area energy efficiency of massive MIMO cellular systems. IEEE Trans. Veh. Technol. **65**(5), 3243–3254 (2016)
16. Television station, December 2017. https://en.wikipedia.org/wiki/Television_station
17. Optimization toolbox, December 2017. https://www.mathworks.com/products/optimization.html

Opportunistic Fog Computing for 5G Radio Access Networks: A Position Paper

Jofina Jijin and Boon-Chong Seet[(⊠)]

Department of Electrical and Electronic Engineering,
Auckland University of Technology, Auckland 1010, New Zealand
{jofina.jijin, boon-chong.seet}@aut.ac.nz

Abstract. Fog-based radio access networks (F-RAN) are posed to play a pivotal role in the much-anticipated 5th Generation (5G) cellular networks. The philosophy of F-RAN is to harness the distributed resources of collaborative edge devices to deliver localized RAN services to the end users. The current F-RAN is implemented mainly utilizing dedicated hardware and do not leverage on the available large number of distributed edge devices. This paper introduces the idea of opportunistic fog RAN (OF-RAN) which comprises of virtual fog access points (v-FAPs). The v-FAPs are formed opportunistically by one or more local edge devices also referred to as service nodes, such as WiFi access points, femtocell base stations and more resource rich end user devices under the coverage and management of the physical FAP, which can be dedicated fog server, fog-enabled remote radio heads (RRHs) or macrocell base stations. The proposed OF-RAN can be a low latency and high scalable solution for 5G cellular networks.

Keywords: Opportunistic fog computing · 5G radio access network
Virtual fog access points

1 Introduction

The future 5th Generation (5G) cellular networks will not only cater high-speed and reliable human communication services, but also support communications between a large number of smart objects or 'things' in the coming era of the Internet of Things (IoT) [1]. To sustain these objectives, centralized radio access networks (C-RAN) was developed where client data received by base stations (BSs) are transmitted over fiber links to a central unit for processing on specialized hardware [2]. Recently, the concept of Cloud-RAN was proposed to replace the specialized hardware in C-RAN with commodity cloud-computing platform to allow for more flexible splitting and allocation of RAN functionalities between radio access points (RAPs) and the cloud, depending on the available cloud resources. However, Cloud-RAN has the shortcomings of: (i) constrained backhaul capacity; (ii) load concentration on the centralized base band unit (BBU) pool; and (iii) difficulty in meeting the ultra low-latency requirements of 5G [3].

More recently, Fog-computing based Radio Access Network (F-RAN) is proposed as a promising candidate to tackle the aforementioned challenges. Fog computing is a

© ICST Institute for Computer Sciences, Social Informatics and Telecommunications Engineering 2018
P. H. J. Chong et al. (Eds.): SmartGIFT 2018, LNICST 245, pp. 82–92, 2018.
https://doi.org/10.1007/978-3-319-94965-9_9

paradigm that extends cloud computing by placing cloud-equivalent resources including processing and storage resources at the edge of the network [4]. In literature, fog computing is also considered as a more general concept of mobile edge computing (MEC). F-RAN harnesses the presence of such collaborative edge devices to deliver localized RAN services to end-users [5]. Its main philosophy is to make full use of local radio signal processing, cooperative radio resource management (CRRM) and distributed storage capabilities in edge devices [6]. Through ingestion and processing of end-user tasks close to their sources, F-RAN has potential to meet the stringent latency and bandwidth requirements of 5G services and applications.

However, the current proposed fog access points (FAPs) of F-RAN have been implemented mainly as dedicated fog servers, or fog-enabled remote radio heads (RRHs) or macrocell base stations [7]. It does not leverage on the presence of a large number of other distributed edge devices in the proximity of FAPs such as WiFi access points, femtocell base stations, and resource-rich end-user devices that can be incentivised to lease their resources and collaboratively serve as 'service' nodes to other end-users. This motivates us to propose the *opportunistic fog RAN (OF-RAN)* inspired by the concept of opportunistic resource utilisation network (oppnet) [8].

The oppnet is a type of specialised ad hoc network that features opportunistic expansion and opportunistic utilization of local resources gained by the expansion. It is a dynamic form of network comprising originally of a small set of 'seed' nodes, which can be expanded on-demand by recruiting 'helper' nodes in their local areas not employed initially but join the seed nodes in order to fulfill a given task. The FAPs in the current F-RAN and the local edge devices (also referred to as *service nodes* in our proposed OF-RAN) can be considered to resemble the seed nodes, and helper nodes, respectively, in oppnet [8].

Hence, this paper also proposes the concept of a *virtual FAPs (v-FAP)* formed by two or more local edge devices and monitored by physical FAPs for 5G radio signal processing. Intuitively, by being in close proximity to the user equipment (UE) that generate the processing tasks and harnessing the collective plethora of local computing resources, the proposed OF-RAN could emerge as a low-latency and high-scalability alternative to the current F-RAN and Cloud-RAN approaches.

The rest of the paper is organized as follows. Section 2 reviews the research literature on investigating various issues of Cloud-RAN and F-RAN. Section 3 presents the proposed OF-RAN architecture. The knowledge gaps are identified and corresponding research questions are discussed in Sect. 4. Finally, Sect. 5 concludes the paper.

2 Literature Review

This section performs a critical review of the relevant literature according to the type of research problems addressed. Four main research problems are identified, namely: (i) limited fronthaul capacity; (ii) flexibility and cost of deployment; (iii) user access mode selection; and (iv) service node selection.

2.1 Limited Fronthaul Capacity

In [11], the authors addressed the limited capacity of fronthaul links in Cloud-RAN by proposing an adaptive compression and joint detection scheme at baseband unit (BBU) pool, which exploits the correlation among the remote radio heads (RRHs) to minimize the fronthaul transmission rate while satisfying the quality of service (QoS) requirements. The RRHs are less sophisticated compared to classical base stations, thus they are considered as relaying nodes that forward IQ signals from the user equipment (UE) to the BBU pool. The block error rate (BLER) of the proposed scheme is analyzed in closed form by using pair wise error probability (PEP). Analytical result showed that a compression efficiency of 350% can be achieved by the proposed scheme. However, the authors have assumed that the BBU always have perfect knowledge of all channel information, which may not be practical and can incur significant overheads in their acquisition due to frequent large-scale message exchanges. When it comes to limited capacity of front haul, overhead reduction becomes more important in Cloud-RAN, which is intended to support a large number of users.

In [12], a joint power control and fronthaul rate allocation scheme for uplink communication in an OFDMA based cloud-RAN is proposed. The proposed scheme is designed for throughput maximization under fronthaul capacity constraint, which is found to have a significant impact on the optimal power control policy. The result showed that the joint design approach achieved better performance than an approach based on optimizing only power control or fronthaul rate allocation. However, the authors have assumed all fronthaul links to be of equal rates and perfectly lossless, and all mobile users are pre-allocated with the same number of sub-carriers, which made the analysis simple, but may not be realistic in real-world heterogeneous environment.

The authors in [13] studied the joint design of cloud and edge processing for the downlink in F-RAN. The BBU performed joint processing for its enhanced RRHs (eRRHs), which cache frequently requested contents of their users, in addition to being a conventional RRH. The objective was to maximize the minimum delivery rate of requested files under the constraints of limited fronthaul capacity and eRRH power. Two fronthauling modes: hard and soft transfer, with different baseline and pre-fetching strategies are considered. In hard transfer, non-cached files are delivered over the fronthaul links, whereas in soft transfer, fronthaul links conveyed quantized baseband signals as in a cloud-RAN. A simulation performance comparison between hard and soft transfer showed that the latter is more effective in using fronthaul resources except in very low signal-to-noise ratio (SNR) regime. However, the authors have largely adopted a heuristic approach to associating UEs to eRRHs, which may not achieve optimal performance as would be achieved under a more theoretically-grounded approach, e.g. by formulating a user association problem which finds the optimal set of eRRHs to associate with the UEs such that the minimum delivery rate is maximized under the constraints of limited fronthaul capacity and eRRH power. Furthermore, the non-associated eRRHs may be put into sleep mode in order to reduce the energy expenditure.

In [9, 10], the authors investigated the performance of a Cloud RAN under flexible centralization, which refers to the concept of suitably proportioning the BBU processing chain (or functional split) between the cloud and RRH, in order to alleviate the issue of limited fronthaul capacity. Various centralization options are analyzed with respect to their required fronthaul capacity, achievable latency and challenges for the signal processing. To enable different information types beyond raw I/Q samples to be transported over the fronthaul when a functional split is implemented between BBU and RRHs, a packet-based transport approach is proposed in [10]. However, this requires a careful design of the packetization method in order to minimize both header-related overhead and payload-filling latency. It is generally observed that existing Cloud RANs have difficulties in keeping their latencies within the timing requirements of the Long-Term Evolution (LTE) standard [10]. More specifically, the overall processing has to be completed in 3 ms in order to comply with the Hybrid Automatic Repeat Request (HARQ) timing, which is the most critical timing requirement defined in LTE [18].

2.2 Flexible and Low-Cost Deployment

In [14], the authors investigated millimeter-wave (mmWave) downlink transmission for the ultra-dense cloud radio access network (UD-CRAN). The fronthaul is shared among RRHs via time division multiple access (TDMA). The joint resource allocation over TDMA based mmWave fronthaul and orthogonal frequency division multiple access (OFDMA) based wireless transmission is studied to maximize the weighted sum rate of all users. The authors have specifically considered a system, where user assigned on any sub-carrier frequency can potentially be served by multiple RRHs subject to fronthaul constraint. The numerical solutions showed that the proposed solution for OFDMA based UD-CRAN can achieve throughput gains of more than 150% over a conventional LTE-A where each user is associated with a single RRH and the mmWave fronthaul bandwidth is equally divided among RRHs. However, the authors have assumed a clear line-of-sight for the mmWave link between RRH and BBU, which may not be possible in densely urban or hilly terrain environments.

The authors in [15] proposed a low cost approach to network densification through on demand deployment of mobile small cells using either mobile handsets or remote radio units. The mobile small cell base stations transmit RF signals to UE in downlink or forward baseband signals from UE to BBU pool for further processing in the uplink. The simulation result showed that proposed approach improved throughput and service quality over the coverage of the network. The proposed solution does not require extensive network planning, but there is high potential for inter-cell interferences with the deployment of heterogeneous small cells alongside macro-cells.

2.3 User Access Mode Selection

The authors in [16] proposed an adaptive algorithm for downlink F-RAN users to select between two content access modes: fog access point (FAP) and device-to-device (D2D), by taking into consideration of their locations, cache sizes and fronthaul delay cost. The proposed algorithm is based on the evolutionary game approach and comprises of three entities: players, strategies, and payoff. Players are users who can choose between multiple access modes. Strategies refer to the selection method, and payoff quantifies the performance satisfaction level of a potential player. Simulation results showed that the proposed scheme can achieve better payoffs than a maximum rate algorithm. However, the author have not considered the channel conditions between the FAP, F-RAN and D2D users in their proposed user access mode selection.

The authors in [17] proposed a centralized opportunistic access control (COAC) with user access mode selection. They considered a D2D underlaid cellular network composed of both D2D user equipment (DUE) and cellular user equipment (CUE), i.e. DUEs communicate with each other using the same radio resources as the CUEs. The user access mode selection, i.e. for selecting between cellular or D2D mode, is based on the user's signal-to-interference ratio (SIR) with respect to cellular base station and DUEs, and the achievable spectrum efficiency. The COAC scheme is compared with a distributed random access control scheme (DRAC) where sub-channels are allocated randomly. The simulation results showed that the user access mode for COAC performed better than the DRAC scheme. However, little attention has been given to the study of considering fronthaul delay in the user access mode selection and its impact on the network latency performance.

2.4   Service Node Selection

The authors in [18] focused on achieving ultra-low latency in F-RAN and proposed an algorithm to determine the optimal number of F-RAN nodes (small- and macro-cell BSs) and amount of resources required for a given distributed computing task. The optimisation problem is firstly formulated to tackle the trade-off between communication and computing resources, followed by cooperative task computing to decide how many F-RAN nodes should be selected with proper resource allocation and computing task assignment. Under the proposed scheme, a target user first sends its processing data to a nearby master F-RAN, which then selects an F-RAN node to serve the user, and is responsible for splitting and combining the tasks. Simulation results showed that the proposed scheme can significantly reduce the total service latency and achieve ultra-low latency. However, the authors have not considered the pre-existing computing load of the F-RAN nodes, as well as the load balancing issue, when assigning a new task to them.

The authors in [19] investigated the formation of a femto-cloud (coalition of femtocell access points) for collaborative processing, in order to avoid using remote cloud while enhancing user's quality of experience (QoE). A cooperative game approach to forming the femto-cloud is proposed, such that the available computation resources are maximally exploited while participating femtocell access points are selected based on satisfying the user's quality of experience (QoE) and monetarily rewarded in a fair manner. The femto cell manager (FCM), which is installed and maintained by the network operator coordinates the formation of femto-cloud. The FCM is also responsible for facilitating information exchanging with neighbouring FCMs. The simulation results shows that the execution delay by using femto-cloud can be reduced up to 50% when compared to that by a single femtocell access point. However, very little attention has been given to the load distribution among the femtocell access points.

In [20], the authors proposed an algorithm for selecting small-cell BSs in a small-cell cloud (similar to femto-cloud in [19]) to process offloaded applications from

Table 1. Summary of literature review

Issues investigated	Proposed solutions	Ref.
Limited capacity fronthaul	• Aim: Joint detection and decompression algorithm to minimize the fronthaul transmission rate • Result: Compression efficiency of 350% can be achieved by proposed optimization schemes	[11]
	• Aim: Joint wireless power control and fronthaul rate allocation optimization in the uplink communication of an OFDMA based C-RAN to maximize the network throughput • Results: Joint design achieves significant performance gain compared to optimizing either wireless power control or fronthaul rate allocation	[12]
	• Aim: Joint design of cloud and edge in order to maximize the delivery rate of the requested files while satisfying the fronthaul capacity • Results: Soft transfer mode provides more effective way of fronthaul resources compared to hard transfer mode	[13]
	• Aim: Functional split on the fronthaul capacity and the use of packet-based fronthaul network • Results: Split of functionality can improve significant centralization gain compared to distributed detection methods	[10]
Flexible and low cost deployment	• Aim: Joint resource allocation for both TDMA and OFDMA based UD-CRAN to maximize weighted sum rate of all users using mmWave fronthaul • Results: OFDMA based UD-CRAN can achieve throughput gain of more than 150%	[14]
	• Aim: Low cost on-demand deployment of mobile small cells using either mobile handsets or remote radio units (RRU) • Results: The proposed solutions improved system capacity and service consistency over the coverage of mobile networks	[15]
User access mode selection	• Aim: Evolutionary game approach for user access mode in fog radio access network • Results: The proposed algorithm has a better user satisfaction than the maximum rate algorithm	[16]
	• Aim: Opportunistic selection of sub-channels based on centralized management by cellular base station, and the criteria for selection between D2D and cellular access mode • Results: D2D mode is a suitable candidate for closely located UEs with strong LOS. But as number of D2D links increases, cellular mode is selected to reduce interference among D2D links	[17]
Service node selection	• Aim: An algorithm to determine optimal number of F-RAN nodes and amount of distributed resources for a given computing scenario • Results: The proposed algorithm can significantly reduce the total service latency of the cooperative task computing operation	[18]
	• Aim: Formation of femto-cloud by local femto cell access points (FAPs) equipped with processing power in a UMTS LTE network via a cooperative game theoretic scheme • Results: The proposed scheme shows superior performance in terms of handling latency and incentives provided to FAP owners	[19]
	• Aim: An algorithm for selection of small cell base station to form the small cell cloud (SCC) to process offloaded data by UEs • Results: The proposed algorithm can achieve 100% user satisfaction as long as the task offloading rate is within a certain limit	[20]
	• Aim: An algorithm for application placement problem with the goal of minimizing the maximum resource utilization • Results: The theoretical and simulation results show the proposed algorithm can balance the computation load well	[21]

UEs. The algorithm considers both UE's computation demand and the computation capacity and load of small-cell BSs in order to achieve high user QoS while maintaining relatively balanced communication and computation load among small-cell BSs. The simulation results showed that the proposed algorithm can achieve 100% user satisfaction as long as the task offloading rate is within a certain limit. However, the authors have not considered the dynamicity of the network, such as UE mobility and changing available computation capability during the processing of offloaded application.

The placement of decomposable application components onto physical MEC nodes was investigated in [21]. The user application and physical nodes are modelled as graphs whose nodes and edges represent the computation and communication resource entities, respectively. Several algorithms for placing the application to physical graphs in different scenarios are proposed with the aim of balancing the load and minimizing the sum resource utilisation at the physical nodes. However, the existing work is mainly focused on offloading or placement of application computation to base-station type nodes in MEC. On the other hand, our work addresses the RAN task assignment problem to a virtual group of co-located edge devices, including UEs in F-RAN. A summary of the literature review performed above is given in Table 1.

3   OF-RAN Architecture

The proposed OF-RAN aims to harness the approach of oppnet for opportunistic formation of fogs as local RAN service groups. The concept of v-FAPs is further introduced in which an opportunistic fog is formed by two or more collaborative and local edge devices under the coverage and management of the physical FAPs. Figure 1 shows the proposed OF-RAN architecture with v-FAPs at the access layer.

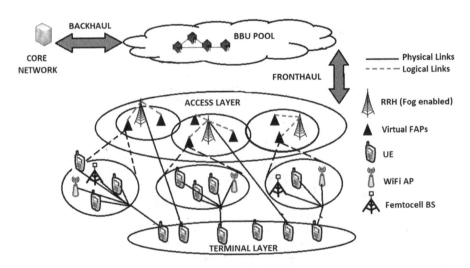

Fig. 1. OF-RAN architecture

As shown in Fig. 1, end-user UEs (hereinafter referred to as client nodes) requests for help from its nearby FAP, which in turn dynamically forms a v-FAP from a set of service nodes (local edge devices) in the client's locality. The FAP decides the set of service nodes to serve the client and the workload assignment to each of the service nodes based on their resource availability. We consider heterogeneous service nodes exhibiting different resource capacities (thus having different costs of utilising resources). The following resource types: computation, storage, communication and energy resources, can be considered when assigning tasks from client node to each service node.

Although the proposed OF-RAN is promising, its performance may be limited by the service nodes that constitute the v-FAPs, which are generally less resourceful than the FAPs or the core cloud. Assigning tasks to service nodes whose resource capacities do not meet the resource requirements of the tasks will result in processing failures. Hence, an important issue is about deciding which service node(s) should be assigned to process what tasks from client nodes in the OF-RAN. The tasks here refer to the client node's radio signal processing tasks that would have normally performed by the FAP or remotely by the cloud. Other important issues that may also need to be investigated include the formation of v-FAP and selection of user access mode (between proposed RAN and conventional RAN).

4   Identified Gaps and Research Questions

In current literature, difficulties have been observed for centralized cloud RANs to keep its latency within timing requirements of the Long-Term Evolution (LTE) standard, and there is still much room for improvement in current F-RAN solutions to effectively address the issues related to cloud RAN. To achieve the best possible outcomes, the following research gaps in both cloud RAN and F-RAN should be tackled.

The capacity limited fronthaul is one of the main challenges faced by current cloud RAN. Although many compression techniques and flexible centralization have been introduced, this requires RRHs to be more sophisticated and capable of handling exponential traffic growth.

The overall processing must be completed in 3 ms in order to comply with the Hybrid Automatic Repeat Request (HARQ), the most critical timing requirement defined in LTE [5]. This tight timing constraint is posing a significant challenge to current cloud-based execution of high-complexity tasks in standard RAN functions such as physical-layer forward error correction (FEC). The excessive load concentration in BBU pool is another issue, as the devices depending on BBU pool for processing are not limited to cellular devices, but also the massive number of Internet of Things (IoT) devices. Current F-RAN mainly comprises of dedicated fog servers, fog-enabled RRHs or macrocell base stations as fog access points (FAPs). Very little attention has been given to the opportunistic expansion of the FAPs through the utilization of a large number of other distributed edge devices such as WiFi access points, femtocell base stations, and resource-rich end-user devices that can be incentivised to lease their resources and collaboratively serve as 'service' nodes to other end-users.

Although the FAPs in the current F-RAN and service nodes in our proposed OF-RAN with v-FAPs resemble the seed nodes, and helper nodes, respectively, in oppnet, but the process of selecting service nodes (recruiting helper nodes) and forming oppnets (v-FAPs) need to be reinvestigated in the specific context of F-RAN. Given differences in the targeted use (e.g. mission-critical application versus RAN service provisioning), type of processing (e.g. application data processing versus radio signal processing), performance requirements, among others need to be considered.

Furthermore, while the OF-RAN seems promising, are there circumstances where the user will still prefer to access the cloud rather than the v-FAP for processing? If so, what the conditions for the user to switch access between the two? The idea of flexible centralization or functional split has so far been only discussed for Cloud RAN, and not for F-RAN. Thus, there is a need to further investigate this concept for F-RAN, particularly under the proposed OF-RAN with v-FAPs architecture. Based on the identified gaps some of the key research question includes:

i. How to efficiently initiate the opportunistic formation of v-FAP and notify the end user? How will information be exchanged between service nodes and seed nodes during the formation process?
ii. What selection criteria should be used for recruiting service nodes in the proposed F-RAN with v-FAPs architecture taking into consideration of finite processing and storage capacity of service nodes, limited energy of battery-operated mobile fog nodes, and 5G signal processing requirements?
iii. What is a feasible way of splitting the radio signal processing function between the BBU and RRHs in an OF-RAN? How can some of these tasks be offloaded to v-FAPs in order to alleviate the burden of RRH and fronthaul?
iv. How should end users (mobile and IoT devices) distribute the 5G signal processing tasks to the v-FAPs, and how will the fog-enabled RRHs collate the processed results of the tasks offloaded to the v-FAPs by the end user?
v. Based on the proposed OF-RAN system model, how to devise a suitable user access mode selection scheme based on both user requirements and available resources?

5 Conclusion

The concepts of Cloud-RAN and F-RAN seems to be very promising for addressing the impact of the exponential growth of the users on the radio access network architecture. However, existing Cloud-RAN and F-RAN solutions are faced by various challenges such as limited fronthaul capacity, load concentration on the centralized BBU, ultra low-latency requirements of 5G and inefficient utilization of the large number of distributed edge devices. In addition, Cloud-RAN may not be suitable for latency sensitive applications.

This paper firstly performed a critical review of relevant literature based on the type of research problems addressed for the existing RANs and their potential solutions. The concept of OF-RAN that features opportunistic expansion and utilization of local edge devices for forming virtual fog access points (v-FAPs) is then proposed and discussed.

Due to the diversity in the multitude of edge devices with respect to their resource capacity and processing capability when it comes to selecting service nodes, a number of important issues need to be investigated. These include deciding which service node(s) should be assigned to process what tasks from end-users in the F-RAN, the formation of v-FAPs, and the selection of user access mode (between OF-RAN and conventional RAN).

The outcomes of this research can benefit both operators and end-users of 5G networks: operator can offload its processing load from the cloud to v-FAPs for more scalable operations; end-users can experience lower latency when served locally by fog devices than remotely by the cloud, thus enhances the provisioning of real-time networked services.

References

1. Tang, J., Tay, W.P., Quek, T.Q., Liang, B.: System cost minimization in cloud RAN with limited fronthaul capacity. IEEE Trans. Wirel. Commun. **16**, 3371–3384 (2017)
2. Peng, M., Sun, Y., Li, X., Mao, Z., Wang, C.: Recent advances in cloud radio access network system architecture, key techniques, and issues. IEEE Commun. Surv. Tutor. **18**(3), 2282–2308 (2016)
3. Checko, A., Christiansen, H., Yan, Y., Scolari, L., Kardaras, G., Berger, M., Dittmann, L.: Cloud RAN for mobile networks - a technology overview. IEEE Commun. Surv. Tutor. **17** (1), 405–426 (2015)
4. Peng, M., Yan, S., Zhang, K., Wang, C.: Fog computing based radio access networks: issues and challenges. IEEE Netw. Mag. **30**(4), 46–53 (2016)
5. Shih, Y., et al.: Enabling low-latency applications in fog radio access networks. IEEE Netw. **31**(1), 52–58 (2017)
6. Hung, S., et al.: Architecture harmonization between cloud radio access networks and fog networks. IEEE Access **3**, 3019–3034 (2015)
7. Dastjerdi, A.V., Gupta, H., Calheiros, R.N., Ghosh, S.K., Buyya, R.: Fog computing principles architecture and applications, Chap. 4. In: Buyya, R., Dastjerdi, A.V. (eds.) Internet of Things: Principles and Paradigms. Elsevier, Massachusetts (2016)
8. Lilien, L., Gupta, A., Kamal, Z., Yang, Z.: Opportunistic resource utilization networks—a new paradigm for specialized ad hoc networks. Comput. Electr. Eng. **36**(2), 328–340 (2010)
9. Wubben, D., Rost, P., Barlett, J., Lalam, M., Savin, V., Gorgogolione, M., Dekorsy, A., Fettweis, G.: Benefits and impact of cloud computing on 5G signal processing. IEEE Sig. Process. Mag. **31**(6), 35–44 (2014)
10. Chang, C.-Y., Schiavi, R., Nikaein, N., Spyropoulos, T., Bonnet, C.: Impact of packetization and functional split on C-RAN fronthaul performance. In: Proceedings of IEEE International Conference on Communications, ICC, Kuala Lumpur, Malaysia, May 2016
11. Vu, T.X., Nguyen, H.D., Quek, T.Q.: Adaptive compression and joint detection for fronthaul uplinks in cloud radio access networks. IEEE Trans. Commun. **63**(11), 4565–4575 (2015)
12. Liu, L., Bi, S., Zhang, R.: Joint power control and fronthaul rate allocation for throughput maximization in OFDMA-based cloud radio access network. IEEE Trans. Commun. **63**(11), 4097–4110 (2015)
13. Park, S.-H., Simeone, O., Shitz, S.S.: Joint optimization of cloud and edge processing for fog radio access networks. IEEE Trans. Wirel. Commun. **15**(11), 7621–7632 (2016)

14. Stephen, R.G., Zhang, R.: Joint millimeter-wave fronthaul and OFDMA resource allocation in ultra-dense CRAN. IEEE Trans. Commun. **65**(3), 1411–1423 (2017)

15. Radwan, A., Huq, K.M.S., Mumtaz, S., Tsang, K.-F., Rodriguez, J.: Low-cost on-demand C-RAN based mobile small-cells. IEEE Access **4**, 2331–2339 (2016)

16. Yan, S., Peng, M., Abana, M.A., Wang, W.: An evolutionary game for user access mode selection in fog radio access networks. IEEE Access **5**, 2200–2210 (2017)

17. Peng, M., Li, Y., Quek, T.Q., Wang, C.: Device-to-device underlaid cellular networks under Rician fading channels. IEEE Trans. Wirel. Commun. **13**(8), 4247–4259 (2014)

18. Chiu, T.-C., Chung, W.-H., Pang, A.-C., Yu, Y.-J., Yen, P.-H.: Ultra-low latency service provision in 5G fog-radio access networks. In: Proceedings of IEEE 27th Annual International Symposium on Personal, Indoor, and Mobile Radio Communications, PIMRC, Valencia, Spain, September 2016

19. Tanzil, S.S., Gharehshiran, O.N., Krishnamurthy, V.: Femto-cloud formation: a coalitional game-theoretic approach. In: Proceedings of IEEE Global Communications Conference, GLOBECOM, San Diego, CA, USA, December 2015

20. Vondra, M., Becvar, Z.: QoS-ensuring distribution of computation load among cloud-enabled small cells. In: Proceedings of IEEE 3rd International Conference on Cloud Networking, CloudNet, Luxembourg, October 2014

21. Wang, S., Zafer, M., Leung, K.K.: Online placement of multi-component applications in edge computing environments. IEEE Access **5**, 2514–2533 (2017)

A Sustainable Connectivity Model of the Internet Access Technologies in Rural and Low-Income Areas

Maria Elena Villapol[1]([⊠]), William Liu[1], Jairo Gutierrez[1],
Junaid Qadir[2], Steven Gordon[3], Jin Tan[4], Luca Chiaraviglio[5],
Jinsong Wu[6], and Wenjun Zhang[1]

[1] Auckland University of Technology, Auckland, New Zealand
{maria.villapol,william.liu,jairo.gutierrez,
wzhang}@aut.ac.nz
[2] Information Technology University, Lahore, Pakistan
junaid.qadir@itu.edu.pk
[3] Central Queensland University, Rockhampton, Australia
s.d.gordon@cqu.edu.au
[4] China Jiliang University, Hangzhou, China
tanjin@cjlu.edu.cn
[5] University of Rome Tor Vergata, Rome, Italy
luca.chiaraviglio@gmail.com
[6] Universidad de Chile, Santiago, Chile
wujs@ieee.org

Abstract. The Internet has evolved as a critical booster for the economic, social and technical development of human society. Almost half of the world's population is unfortunately missing out due to the lack of access to the Internet. Such users are mainly those living in rural and low-income areas. Various strategies and approaches for improving the Internet's accessibility are available, each with a different set of benefits, costs, and risks. It is important to choose solutions from these feasible options that promise to promote the efficiency as well as the sustainability of the 'Internet Ecosystem'. In this paper, we propose a new model of *sustainable connectivity* that integrates three factors (affordability, social shareability, and geographical network coverage) that must be considered in the selection and design of Internet access solutions. In addition, we develop a hypergraph-based network graph solution that illustrates the relationship among the three factors. Then, we use Coloured Petri Nets (CPNs) to model and simulate the possible Internet access solutions and also interplay those three factors to study how they impact the overall network connectivity performance. Our initial results have revealed how sustainable Internet connectivity behaves as a function of the affordability, social interaction, and geographical network coverage and investigates how these factors could be leveraged to provide different network connectivity and Internet access solutions.

Keywords: Internet access · Sustainable connectivity
Rural and low-income areas · Affordability · Social shareability
Geographical network coverage · Hypergraph theory · Coloured Petri Nets

© ICST Institute for Computer Sciences, Social Informatics and Telecommunications Engineering 2018
P. H. J. Chong et al. (Eds.): SmartGIFT 2018, LNICST 245, pp. 93–102, 2018.
https://doi.org/10.1007/978-3-319-94965-9_10

1   Introduction

The Internet is vital for a nation's development and its social and economic growth. An open, secure, trustworthy, and universally accessible Internet can facilitate greatly in attaining the United Nations defined Sustainable Development Goals (SDGs) [1]. The 17 goals include ending poverty, protecting the planet, and guaranteeing prosperity to all humankind. Although Internet penetration rates are high (over 80%) in developed countries [2], the people living in rural and low-income areas generally face access problems with limited or non-existent connectivity (60% of offline population live in rural areas [3]). Internet connectivity in these areas is challenging because of barriers such as remoteness of hard-to-reach rural areas, low density of users, and low-income of users.

Internet access in New Zealand is facing similar challenges [4], and the New Zealand Government has launched several initiatives to extend Internet connectivity in rural areas [4]. One of them is the Rural Broadband Initiative (RBI), a partnership among the Government and different providers to deploy broadband solutions in rural areas. Despite RBI's progress, rural users are still demanding higher and higher data rates, and more reliable and affordable Internet access. According to the participants in the Rural Connectivity Symposium [5], there is an urgent need to deploy Internet services and networks in order to support health services, such as emergency healthcare services, and to have different connectivity options given that there will be no one-size-fits-all solution. The discussions in the symposium also highlighted the priorities for improving rural connectivity, which include the identification of opportunities for boosting rural economic activity and productivity levels.

Harrison *et al.* [6] have defined the factors which have some impact in the deployment of connectivity infrastructure initiatives and projects, such as: the availability and quality of the Information and Communication Technology (ICT) infrastructure, the accessibility to the Internet for education, communication and health services, and affordability as some of the digital divide indicators [7]. These factors still hold in rural cases.

In this paper, we propose a sustainable connectivity model for rural and low-income areas, in order to provide Internet access. The model is based on the following pillars: affordability, social shareability and geographical network coverage. Based on these pillars, we then analyse the reachability of users to identify network access technologies suitable for the rural and remote users. More formally, the reachability of users for each individual factor can be represented as a graph. We then use a hypergraph to model the relationship between the different features, where each dimension of the hypergraph refers to one reachability area. In this way, we are able to leverage on the hypergraph theory [8, 9] to identify the optimal access technologies. Moreover, we use Coloured Petri Nets (CPNs) [10] to demonstrate the applicability of our proposed model. In particular, the outcome is a set of results showing the percentage of suitable connections for several optimistic and pessimistic assumptions. Employing CPNs to simulate different scenarios has the advantage of offering initial insights into the model effectiveness at a high level of abstraction, while still being able to include detailed scenario models and to obtain more realistic results in the future.

The paper is organized as follows. Section 2 provides an overview of the affordability, social and geographical critical factors with special emphasis in the New Zealand context. Section 3 then presents a sustainable connectivity model for Internet access in rural and low-income areas. Section 4 then reports the results, which are obtained by applying CPNs on the considered case study. Finally, Sect. 5 concludes the paper.

2   Related Work

In this section we review the literature by identifying the importance of considering the reachability of users based on these factors.

Social shareability represents the willingness of users to share their network connections so other users can gain access to the Internet through the shared connection. The idea of sharing the Internet connection for social purposes has already been proposed as a solution for providing connectivity to low-income families living in an urban area in [11]. Vural et al. [12] identified the sharing of broadband connections as one of the attractive options for increasing wireless connectivity in urban areas when deploying wireless mesh networks.

Focusing on the affordability, the UK telecommunications regulator defines this aspect as the capability of a good or service to be purchased by a consumer without suffering undue hardship [13]. Affordability is one of the digital divide factors, meaning that some rural users may not be able to afford Internet access [7]. In this context, the reachability of the affordability means that a user can pay for the connection given the Internet access cost. In particular, the authors in [14] introduce a reachability analysis given the wages of the users.

In [4], several potential approaches for providing broadband connectivity in rural areas of New Zealand are discussed. The authors consider the socio-technical needs of the potential rural users in order to get them engaged in the development of their connectivity solutions. Moreover, a set of four rural access technologies is overviewed. In this context, "geographical reachability" means that a user can reach another user to establish a communication link. The abovementioned technologies provide different network coverage and support different levels of user mobility which need to be taken into account for a geographical reachability analysis. Moreover, Durairajan et al. [15] propose a framework for identifying opportunities for broadband connectivity deployment. The authors consider different factors, including: the infrastructure availability, the user demographics, and the deployment costs. Differently from our work, the complex relationships among the defined factors are not considered. Moreover, both social and affordability factors are not taken into account.

In our work, we adopt hypergraph theory and CPNs. Hypergraph theory is a powerful tool to model complex relationships among objects within a system. For example, Bai et al. [16] propose a hypergraph framework to formulate the complex relationships among the entities in a caching based D2D communication system. Hypergraphs are beneficial for our research as they allow reachability analysis across multiple dimensions (factors) and can be used to find optimal solutions with respect to the selected Internet access technology. Moreover, CPNs have been used extensively to build models of distributed systems at different levels of abstractions, and to obtain numerical simulations results [10]. They are beneficial for our research because we can abstract away from many details and we can easily obtain a set of initial results.

3   A Sustainable Connectivity Model for Rural Zones

The proposed sustainable connectivity model includes three factors: affordability, social shareability and geographical reachability as shown in Fig. 1. It also shows the possible relationships among these three factors associated with the potential access technologies, thus we can evaluate these technologies through a 3D-perspective. Here we denote the x-axis as the geographical reachability, the y-axis as the affordability reachability and the z-axis as the social shareability. For example, the D2D wireless communications technology with mesh networking could be a suitable access solution when the social shareability is high, but the geographical reachability is low. In order to measure the possible technological solutions for selection, we could use this sustainable connectivity model to leverage and to optimize the cooperation among the three variables while keeping fixed the total amount of the resources.

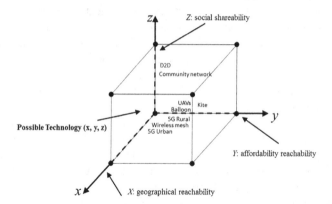

Fig. 1. The three-dimension based sustainable connectivity model.

Figure 1 does not fully capture the model, as the shareability and reachability factors are not singular values but rather complex relationships between users in a particular area. Therefore, the initial sustainable connectivity model needs to be further enhanced by representing it as multiple graphs with different layers. Figure 2 reports a representative example of a 13-node network. In the geographical dimension, the nodes represent the user locations, and the links represent the existing physical connections, e.g., cables or fibres already established between two users by the infrastructure provider. In the affordability dimension, the links between two nodes represent the fact that one user can afford to connect with another. Finally, in the social dimension, each link represents the two users with willingness to allow sharing their devices to connect with each other wirelessly so as to extend the Internet connectivity.

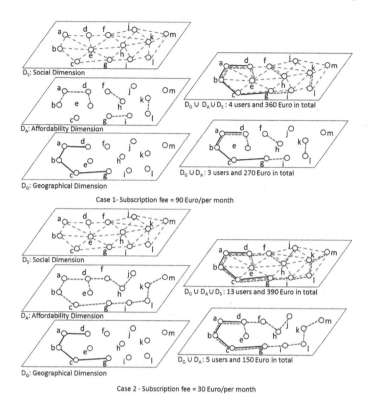

Fig. 2. An example of sustainable connectivity model exploiting multi-dimensional graphs.

We then further consider two representative cases (reported in Fig. 2). In the first one (top part of Fig. 2), we assume that the subscription fee to the Internet service is very high, i.e. 90 [EUR] per month. Firstly, we consider both geographical and affordability factors i.e., $D_G \cup D_A$, for the Internet access. It can be seen that only three users (i.e., a, b and d) can access the Internet while users c and g have network connectivity but they cannot afford it. On the other hand, users (f, h, g, i, k, l) can afford it but they do not have network connectivity yet so they cannot access the Internet either. In rural areas, neighbourhoods in small towns or villages usually tend to cluster people in relatively small areas. Therefore, we can assume denser link connectivity in its social dimension. In other words, the users have willingness to support and share their Internet connectivity to each other if the network security and payment issues can be addressed. In this case, the social shareability factor can be included (i.e., $D_G \cup D_A \cup D_S$) and D2D mesh networking can be set up among users; in this scenario, there is an extra user who can be further connected to access the Internet service.

Focusing on the second case (bottom part of Fig. 2), we reduce the monthly subscription fee from 90 Euro to 30 Euro. As a result, more users can afford it, i.e., more links in the affordability layer are established, while the links in the other two dimensions are kept unchanged. It can be seen that there will be 5 users who can access the Internet when both geographical and affordability factors are considered. This number is increased to 13 users when all three factors are considered. From the network operator's perspective, the option of having more users (i.e., 5 users vs. 3 users) but with lower profit (i.e., 150 Euro vs. 270 Euro) is not a feasible solution. On the other hand, the solution of having more users (i.e., 13 users vs. 3 users) with more profit (i.e., 390 Euro vs. 270 Euro) could be an attractive option to pursue. This could be a win-win solution between the network operators and end-users.

As shown above, the traditional graph approach is not sufficient to holistically model the complex relationships between affordability, geographical reachability and social shareability. In order to explore the interplay among these three factors, we can represent our sustainable connectivity model by using multi-dimensional graphs through the use of hypergraph theory. Hypergraph theory provides the mathematical foundation required to formulate the complicated relationship among these factors. It can also facilitate the understanding of those relationships and allow us to carry out further studies of our proposed model. The hypergraphs are extensions of graphs which can model more general types of relationships [9]. The formal definition of a hypergraph is as follows.

Definition 1. A Hypergraph is a pair $H = (V, E)$, where

1. $V = \{v_1, v_2, v_3, ..., v_n\}$ is the set of vertices or nodes
2. $E = \{E_1, E_2, ..., E_m\}$, E_i is a subset V, for $i = 1, ..., m$, is the set of hyperedges.

If the hypergraph does not have any isolated vertex

$$\bigcup_i^m e_i = V$$

An isolated vertex x is defined as

$$x \in V \backslash \bigcup_i^m e_i$$

A hyperedge e is a loop if $e \in E$ such that $|e| = 1$

In order to study the relationships among the three factors, we propose a hypergraph representation of the sustainable connectivity measurement model. We denote the set of the communication network users as V^c, where x_n^c is the nth user of the communication access network. The set of users which may have a social interaction is denoted by V^s, where x_m^s is the mth user. We denote the set users in the affordability graph as V^a. The set of vertices in the proposed hypergraph is defined as:

$$V^c \cup V^s \cup V^a = V$$

In the considered hypergraph for the sustainable connectivity measurement model, a hyperedge exists if and only if:

- Two users of the given communication network, which belong to V^c, are willing to establish a communication link.

- One of them is interested in sharing the connection with the other user x_m^s.
- One of the users, x_n^a can afford the connection.

The benefit of representing the three factors as a hypergraph is that efficient analysis techniques can then be used to identify optimal solutions. As shown in the examples of Fig. 2, for a target area for deploying Internet access, it is possible to generate hypergraphs for different access technologies, given the users and the geographical data (e.g. income data, social interactions such as phone call frequencies, mobility patterns, and radio signal propagation maps). The analysis of the hypergraphs could reveal which access technologies can provide suitable Internet access to the most users with the least cost.

4   A Case Study Based on Coloured Petri Nets

Having considered the sustainable connectivity model as a hypergraph, we want to better investigate the potential of the proposed three-dimensional solution. In this context, we adopt CPNs as a modelling and simulation tool because they allow the creation of models at different levels of abstraction. Thus, we can generate models that represent the three dimensions of the sustainable factors at a level of abstraction. This model captures the functional properties which need to be proved and allows us to analyse the system despite its intrinsic complexity.

In the following, we describe how the CPNs are exploited. In particular, we exploit the hierarchical constructs of CPNs [10]. The top-level module is shown in Fig. 3, which depicts a three-dimensional representation of the model shown in Fig. 2. This module includes a *substitution transition* (drawn as rectangles) for each reachability dimension, each of them defined by its own module. In the figure, the *place* (drawn as ellipses) named "*Users*" represents the users of the communication access network who may want to interact with other users and may be able to afford the connection. The place named "*SysState*" represents the state of each sub-system, i.e., affordability, social shareability and geographical reachability. Places and transitions are connected by *arcs* which have expressions associated with them.

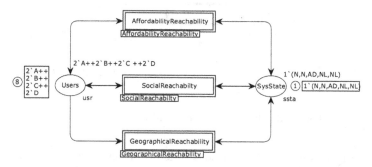

Fig. 3. Top view of the CPN module for the sustainable connectivity model.

We conduct simulations for the scenarios shown in Table 1. Each scenario is defined by three probability variables in the simulation: the probability of affordable reachability (p_a) means that the user can afford a connection with probability p_a; the probability of social shareability (p_s) means that a user has a chance to share the connection with another user with probability p_s; and the probability of geographical reachability (p_g) means that the user can reach another user with probability p_g. The probability values have been chosen to represent either pessimistic scenarios (i.e. scenarios 4, 5, 6, and 8) or optimistic ones (i.e. scenarios 1, 2, 3 and 7). Examples of the scenarios are shown in Fig. 2 and described in Sect. 3.

Table 1. Simulation scenarios

Scenario	Description	Affordability	Social	Geographical
1	Poor geographical reachability	0.9	0.9	0.3
2	Poor social links	0.9	0.3	0.9
3	Poor affordability	0.3	0.9	0.9
4	Poor geographical reach. & social	0.9	0.3	0.3
5	Poor geographical reach. & afford.	0.3	0.9	0.3
6	Poor afford. & social links	0.3	0.3	0.9
7	Optimistic	0.9	0.9	0.9
8	Pessimistic	0.3	0.3	0.3

Figure 4 shows the module for the reachability of an affordable connection (the social shareability and the geographical reachability modules are similar). For the sake of simplicity, we use a uniform distribution to represent the probability that a user can afford a connection. We conduct simulations for a network with four nodes (A, B, C and D), which is shown in Fig. 3. The initial marking (i.e. the initial state of the system) is the initial distribution of tokens to the model places, where a *token* is a value (colour), which belongs to the type of the place.

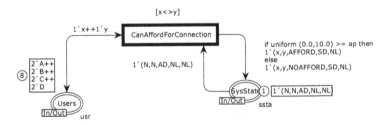

Fig. 4. Reachability module for the affordability factor.

Figure 5 reports the simulation results in terms of the reachability states for scenarios 1–8. A reachability state shows the state of each graph of the system where *Can Afford* means that a user can afford the connection, *Social Shareability* means that the user can (or is willing to) share the connection, and *Network Coverage* means that the user is in the geographical area of the network. The reachability state 4 is the desired state where two users who are interested to share the connection with each other can establish a communication link and pay for the Internet access service (i.e., a hyperedge exists). In the scenarios where two or more of the reachability factors are favourable (i.e., scenarios 1, 2, 3 and 7), we can see that there is a better chance to reach the desired reachability state. On the other hand, if at least two reachability factors are poor, there is a low chance to get the users interacting by exploiting an affordable physical connection. Moreover, the percentage of users who cannot interact because of at least one factor is not met is at most equal to 20% or lower in most of the scenarios.

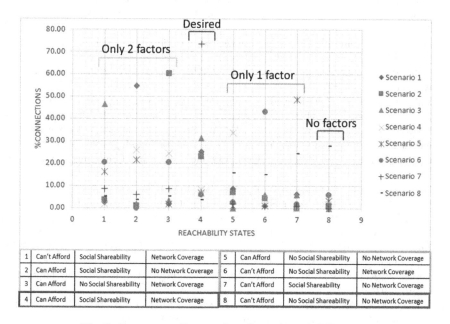

1	Can't Afford	Social Shareability	Network Coverage	5	Can Afford	No Social Shareability	No Network Coverage
2	Can Afford	Social Shareability	No Network Coverage	6	Can't Afford	No Social Shareability	Network Coverage
3	Can Afford	No Social Shareability	Network Coverage	7	Can't Afford	Social Shareability	No Network Coverage
4	Can Afford	Social Shareability	Network Coverage	8	Can't Afford	No Social Shareability	No Network Coverage

Fig. 5. Percentage of connections for each reachability state

5   Conclusions and Future Work

In this paper we have proposed a sustainable connectivity model of the Internet access in rural and low-income areas. Our model takes into account affordability, social shareability and network geographical coverage factors. We represent our solution as a three-dimensional graph by using hypergraph theory. We then use CPNs to model the 3D graphs and to represent the considered factors at different level of abstractions. By exploiting our model, we have provided insights into more detailed information of users such as whether a user likes to share the connection with others or not, with existing physical connectivity or not, as well as whether they can afford the intended connection or not. All of this information is helpful to evaluate different access technologies. We have then conducted a simple, yet representative, simulation study, by taking into account both optimistic and pessimistic scenarios. Our preliminary results confirm the effectiveness and the potential of our model.

As future work, we plan to provide more complex sustainable connectivity models to accurately capture the affordability, social and geographical situations and their dynamics in the rural areas. Moreover, we will perform a tech-economic analysis for comparing various access technologies in real rural and low-income areas to validate the credibility and the scalability of our model.

References

1. United Nations: Sustainable Development Goals: 17 Golas to Transform our World. http://www.un.org/sustainabledevelopment/sustainable-development-goals/#. Accessed 03 Jan 2018
2. Internet World Penetration Rates. http://www.internetworldstats.com/stats.htm. Accessed 15 Dec 2017
3. Philbeck, I.: Connecting the Unconnected: Working Together to Achieve Connect 2020 Agenda Targets (2017)
4. Villapol, M., et al.: Connecting the unconnected 10% of New Zealanders by 2025 : is a MahiTahi approach possible? In: Proceedings of 27th International Telecommunication Networks and Applications Conference, ITNAC, Second International Workshop on Data Intensive Computing and Communications for Sustainable Development (2017)
5. Connecting Rural New Zealand-Insights from the 2017 Rural Connectivity Symposium (2017)
6. Harrison, C., et al.: Foundations for smarter cities. IBM J. Res. Dev. **54**(4), 1–16 (2010)
7. Barzilai-Nahon, K.: Gaps and bits: conceptualizing measurements for digital divide/s. Inf. Soc. **22**(5), 269–278 (2006)
8. Berge, C.: Hypergraphs: Combinatorics of Finite Sets, vol. 45. Elsevier, Amsterdam (1984)
9. Bretto, A.: Hypergraph Theory: An Introduction, vol. 53. Springer, Heidelberg (2013). https://doi.org/10.1007/978-3-319-00080-0
10. Jensen, K., Kristensen, L.M.: Coloured Petri Nets. Springer, Heidelberg (2009). https://doi.org/10.1007/b95112
11. Villapol, M.E., Abreu, D.P., Cordero, C.: Planning a wireless mesh network which takes advantage of the Urban geography of the city. In: Proceedings - 2015 Asia-Pacific Conference on Computer-Aided System Engineering, APCASE 2015 (2015)
12. Vural, S., Wei, D., Moessner, K.: Survey of experimental evaluation studies for wireless mesh network deployments in Urban areas towards ubiquitous Internet. IEEE Commun. Surv. Tutor. **15**, 223–239 (2013)
13. Ofcom: Results of research into consumer views on the importance of communications services and their affordability (2014)
14. Barrantes, R., Galperin, H.: Can the poor afford mobile telephony? Evidence from Latin America. Telecommun. Policy **32**(8), 521–530 (2008)
15. Durairajan, R., Barford, P.: A techno-economic approach for broadband deployment in underserved areas. ACM SIGCOMM Comput. Commun. Rev. **47**(2), 13–18 (2017)
16. Bai, B., Wang, L., Han, Z., Chen, W., Svensson, T.: Caching based socially-aware D2D communications in wireless content delivery networks: a hypergraph framework. IEEE Wirel. Commun. **23**(4), 74–81 (2016)

E-Mobility: Smart Grid and Charging Session of Electric Vehicles

Gabriele Corzato[1], Luca Secco[2], Arslan Rasheed[3],
Atulya Kumar Nagar[4], and Emanuele Lindo Secco[4(✉)]

[1] Department of Industrial Engineering, University of Padova,
Via Venezia, 1, Padua, PD, Italy
gabriele.corzato@studenti.unipd.it
[2] DriWe, Contrà progresso, 1/H, Schio, VI, Italy
luca.secco@driwe.eu
[3] Engineering Research Institute, Auckland University of Technology,
Auckland, New Zealand
arslan.rasheed@aut.ac.nz
[4] Robotics Lab, Department of Mathematics and Computer Science,
Liverpool Hope University, Hope Park, Liverpool L16 9JD, UK
{nagara, seccoe}@hope.ac.uk

Abstract. This research stresses upon the importance of electric mobility in modern age as sales of Electric Vehicles (EV) have crossed one million cars and current number of charging stations are rapidly growing. In this context, we propose a novel smart connection called DriWe between the EV, the charging point and the environment to improve and consolidate the development of smart grids vs the charging enhancement. In DriWe, performance of the monophasic environment is optimized, providing support to the owner of the EV in order to (i) always achieve maximum power, (ii) easily reach the charging facilities through a special application that can be installed in tablet, smartphone or pc, and identify the various charge points (iii) guarantee the absence of electric blackout, thanks to a combination and an interaction of three elements: smart device, an intelligent framework, namely a DriWe cloud server and charge point. A dynamic load control method based on an Application Programming Interface (API) with a 10 s frame rate update, is applied. The program runs within the end-user smart phone, acquires data from DriWe cloud server, and allows to coordinate the recharge column by allowing the Electric Vehicle (EV) supply equipment's control and variation.

Keywords: Electric Vehicles · Smart charging · Dynamic control
Charging station

1 Introduction

Since the beginning of the twenty-first century, the e-mobility market has been aiming for a radical and substantial change of mentality by the community. The fundamental objective of this new vision is the reduction of CO_2 emissions, which are the main cause of global temperature rise. Therefore, the next-generation electric power systems

© ICST Institute for Computer Sciences, Social Informatics and Telecommunications Engineering 2018
P. H. J. Chong et al. (Eds.): SmartGIFT 2018, LNICST 245, pp. 103–110, 2018.
https://doi.org/10.1007/978-3-319-94965-9_11

will integrate these diversified renewable energy resources, storage systems, controllable loads and automated & intelligent management systems [1]. Furthermore, as automated and distributed energy network, the smart grid can be conceived as a two-way flow of electricity and communication, which allows monitoring everything from generation to consumer [2].

Basic support for the evolution of a society based on e-mobility will be given by the support of governments, which, after the Paris agreement on climate change, will have to commit themselves to adopt eco-sustainable solutions to solve environmental problems. The scenario of the planet in which electric cars are moving is gaining more and more attention from the researchers and it is destined to become the indispensable promoter of sustainable mobility in the coming decades. The consumers will not automatically switch to Electric Vehicles (EVs) if the cost trend stays high, if the charging station network is not ready, or if new technologies are not easily usable [3].

The easiest way to recharge EVs is the conductive wired charging, a physical connection between two different parts: car and a domestic plug, or an industrial plug or a charging column. Charging in monophasic environments occurs in alternating current. Nevertheless, according to recent investigations, car and domestic plug are not suitable for recharging vehicles for several hours to a current of 16 A due to frequent over-heating.

In AC conductive charging there are two different charging modes for an EV. Whether one uses the connector supplied with the purchase of the EV, simply connect the two ends of the EVSE [4] (which have two types of electrical sockets) to the vehicle and the domestic plug or to the charging station, respectively. If instead one uses a solution where the cable is permanently connected to the charge point, and then supplied with it, you will have a single free connection terminal that will be inserted in the EV [5].

2 Materials and Methods

2.1 The Problem

In a first instance the main problem of recharging an EV in a domestic or commercial environment can be summarize as follows: how to avoid the blackout of the building during the EV charging session?

In order to develop the DriWe system, we came across a relatively large and daily problem designed for families or small restaurants, shops and businesses that still want to offer better service to their customers and bring a structural benefit. Families are the strong point of this implementation, as the charging columns will be a strategic element in all homes.

More specifically, a DriWe dynamic control architecture has been developed: the dynamic control of the power loads during the charging session of EVs can be modulated in real time, based on the amount of energy available upstream of the plant (Fig. 1). A smart device is therefore interposed, which assumes the task of evaluating the priorities of the loads and communicating through a specific program to the charge point whether and how much to deliver. Communication is performed through an Open Charge Point communication Protocol (OCPP).

Fig. 1. The main functional blocks of the DriWe framework

The smart device must have some essential characteristics, such as reliability, speed of communication, immediate data transmission, small size but above all a low price.

2.2   The Optimization Framework

The proposed framework is based on a block diagrammatic representation, which is quite schematic and easily understandable. The implemented program works by recalling the amount of absorbed loads power in the examined environment: this information is collected through a smart device via an API (Application Programming Interface) which updates every 10 s: (i) a sufficient time to prevent the electric meter from blowing if the contractual power limit is exceeded; (ii) time that avoids a clogging of the communication network. Accordingly, through the initialization of some variables (e.g. the contractual power limit, the monophasic voltage) and a mathematical algorithm checks whether the instantaneous power is higher or lower than the contracted limit power. In case:

- The Current Power is greater than the Contractual Limit Power: we would have an excess power that will be subtracted from the charging power supplied by the charge point, and therefore no electrical blackout.
- The Current Power is less than the Contractual Limit Power: the charge point power supplied can be increased.

Therefore, it is possible to calculate the new current to be supplied by the charge point. This new current must be included within a range of values, a lower limit – namely 6 A, according to the CEI EN 61851-1 - and an upper limit, depending on the specifications and requirements of the charging station. The following configurations may occur:

- The new value of the current is equal to the old one: the configuration parameters are not changed.
- The new value of the current is different from the old one: the algorithm modifies the charging station current through a "Change Configuration" menu, a command given by a special Open Charge Point communication Protocol (OCPP) between the system and charging column. The adopted OCPP protocol, version 1.5, consists of 25 operations: 10 of these operations are initialized by the charging station, whereas the other 15 ones are performed by the central control system [6] (Fig. 2).

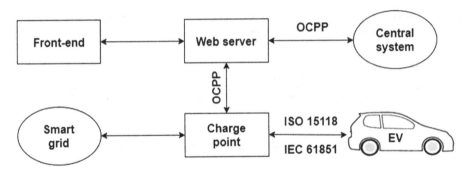

Fig. 2. The DriWe operating scheme [8]

Thanks to these approaches, the configuration parameters of the charging station are changed, maintaining the constant voltage with a variation of the current and therefore of the power delivered to the car. Figure 3 reports an overview of the implementation of the aforementioned process.

3 Experiments and Results

The DriWe has been tested on different commercially available brands and types of EVs. On each test a full charging process has been performed. The tests were carried out at the DriWe laboratories. Each vehicle under examination was monitored for a period of 5 h where it was possible to notice the variation of the current supplied by the charging station vs the changes of the loads occurring in the laboratory. The measurements were performed via the calculation program which has been previously detailed.

We have tested our model on Renault, Twizy and Nissan Leaf to verify the results which are explained below:

Renault Twizy. Figure 4 shows the performed output charging power in [W] vs the recharging time in [s]. The red line refers to the threshold power, namely the value of 3 [kW]. The graph shows how the system started with a simple monitoring of the loads in the environment under examination, then with the start of the recharge phase at time [s] an increase in the power absorbed by the network can be noticed; nevertheless, with the consequent introduction of a further electric load, the total instantaneous power exceeds the simulated contractual power limit of 3 [kW]. The calculation program will then be activated with a consequent reduction in the current supplied to the vehicle.

The last part of the graph shows a series of power peaks due to the recharging phase of the vehicle with its consequent detachment. As it can be observed, the recharge session does not result to be successful because changing the current from management there is a communication error given by the absence of the pilot control, "a control conductor in the power cable connecting the control box on the cable or in the fixed part of the EVSE, and the ground of the EV through the control circuit on the vehicle" [7], which does not allow a correct modulation of the signal and therefore a variation of the current.

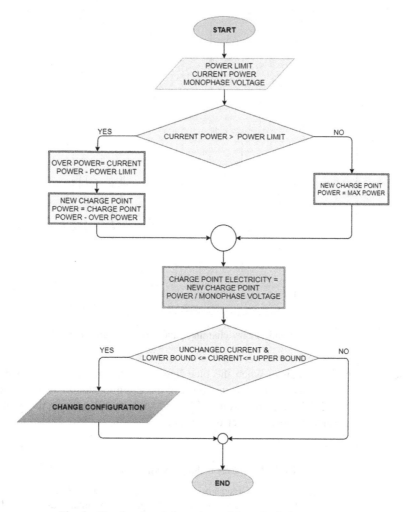

Fig. 3. The functional flow chart of the calculation program

Nissan Leaf. Figure 5 shows the results of the experiment performed on Nissan Leaf in which the first and third part of the graph shows the line of instantaneous absorbed power.

Figure 5 shows that the instantaneous absorbed power - namely the yellow line within the plot of the figure - follows the red line at 4.5 [kW], meaning that the loads plus the charging station are constantly following the contracted power line.

This pattern shows the correct functioning of the calculation program: in fact, when the loads - blue line in the graph - varies within the examined environment and a correct modulation of the current supplied to the vehicle is obtained, always allowing maximum recharging power, avoids exceeding the contractual limit and the risk of

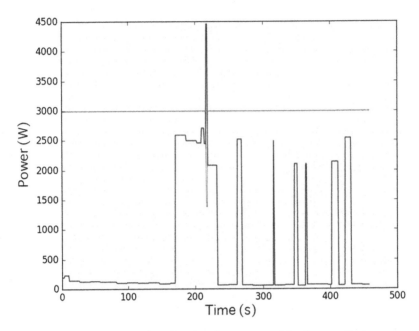

Fig. 4. The Renault Twizy charging pattern (Color figure online)

electric blackout. For instance, when the blue line is overtaken - from 0.75 [kW] to 2.03 [kW] - the current and the maximum power of the charging station decreases from a value of 16 [A] to 10.70 [A] and from 3.68 [kW] to 2.46 [kW], respectively, however the instantaneous absorbed power is constant.

Table 1. Phases recharge Nissan Leaf

Blue line Power loads [kW]	Charging station Current [A]	Green line Maximum power charge point [kW]	Yellow line Instantaneous absorbed power [kW]
0.80	15.60	3.58	3.58 + 0.80 ≈ 4.40
0.75	16.00	3.68	3.68 + 0.75 ≈ 4.40
2.03	10.70	2.46	2.46 + 2.03 ≈ 4.50
1.95	11.10	2.55	2.55 + 1.95 = 4.50
3.90	**6.00**	1.38	1.38 + 3.90 ≈ **5.30**
0.80	16.00	3.68	3.68 + 0.80 ≈ 4.50

By analyzing the central part of the graph, greater absorption of the loads present in the examined environment could be witnessed, making it possible to exceed the contractual threshold up to 5.3 [kW]. The dynamic modulation system, to stay below the 4.5 [kW] threshold, would have to supply about 2.5 [A], but the recharge column has a lower supply limit of at least 6 [A] as it was mentioned before.

The above table (Table 1) reports the main values of the power loads, charging station current and the performed instantaneous and maximum power.

Nevertheless, a threshold overrun has occurred but there is no electric blackout. In order to solve this problem, the following strategies maybe adopted with the aim:

- To implement the calculation program in such a way that if the current supplied by the column is less than 6 [A], the recharge terminal is disconnected: following this approach, all the contractual power is used for the power of the loads of the examined environment.
- To send an alert message (i.e. SMS, e-mail or display) to the user to make him aware of the exceeding of contractual power and the risk of electric blackout if the power of the loads is not reduced.
- To integrate the solution with a home automation device or AAL (Ambient Assisted Living) system with a set load priority.

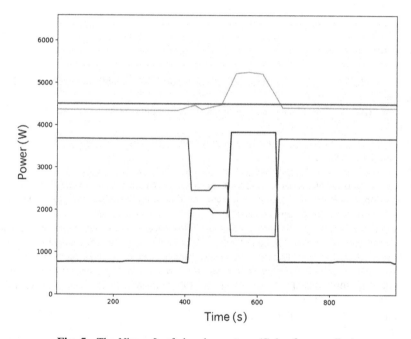

Fig. 5. The Nissan Leaf charging pattern (Color figure online)

4   Conclusions

We propose a novel architecture to improve and simplify the relationship between the end-user and the Electrical Vehicle, making the recharging process more practical [9]. The proposed DriWe system and calculation program allows the monitoring of the overall process every 10 s in terms of power absorbed by the various loads of a reference environment, avoiding electric blackout by varying, within the time limit, the power levels delivered by the charge point towards the EV. The results show that our proposed architecture outperforms the existing architectures. Furthermore, it may be integrated and combined with other smart grid application in the Ambient Assisted Living (AAL) context [10].

A future development consists in interfacing the electricity meter with the charging station through the communication on the appropriate Chain 2 channel. It is therefore necessary to carry out a data transmission by means of the PLC (Power Line Carrier) communication channel, using appropriate waves conveyed on the wiring systems that connects the meter directly to the relative user device, in order to be able to modulate the power supplied by the charging station.

Acknowledgments. We thank you all staff of the DriWe Company for their valuable support and in particular: Mr. Alessio Vitella.

This work was presented in thesis form in fulfilment of the requirements for the M.Sc. in Energy Engineering for the student G. Corzato at the Department of Industrial Engineering, University of Padova, Italy.

References

1. Irena.org (2017). http://www.irena.org/documentdownloads/publications/smart_grids.pdf. Accessed 15 Dec 2017
2. Iea.org (2017). https://www.iea.org/publications/freepublications/publication/smartgrids_roadmap.pdf. Accessed 15 Dec 2017
3. Aci.it (2017). http://www.aci.it/fileadmin/immagini/Notizie/Mobilita/FIA_E-mobility.pdf. Accessed 15 Dec 2017
4. Abronzini, U., et al.: Optimal energy control for smart charging infrastructures with ESS and REG. In: 2016 International Conference on Electrical Systems for Aircraft, Railway, Ship Propulsion and Road Vehicles & International Transportation Electrification Conference, ESARS-ITEC, Toulouse, pp. 1–6 (2016)
5. Fattibilità tecnologica per il controllo dinamico dei carichi in ambiente monofase, durante la ricarica di veicoli elettrici. Gabriele Corzato, Università degli studi di Padova (2017)
6. Rodríguez-Serrano, Á., Torralba, A., Rodríguez-Valencia, E., Tarifa-Galisteo, J.: A communication system from EV to EV Service Provider based on OCPP over a wireless network. In: 39th Annual Conference of the IEEE Industrial Electronics Society, IECON 2013, Vienna, pp. 5434–5438 (2013)
7. Sistema di ricarica conduttiva dei veicoli elettrici, Parte 1: prescrizioni generali. Norma CEI EN 61851-1 (2012)
8. Wellisch, D., Lenz, J., Faschingbauer, A., Pöschl, R., Kunze, S.: Vehicle-to-grid AC charging station: an approach for smart charging development. IFAC, Deggendorf (2017)
9. Secco, L., Alberti, A., Secco, E.L.: MOB-Y: smart grid for sustainable MOBility with retrofit electric vehicles. In: The 3rd IEEE International Conference on Cybernetics, CYBCONF 2017, 21–23 June 2017 (2017). https://doi.org/10.1109/cybconf.2017.7985790
10. Anya, O., Tawfik, H., Secco, E.L.: A context-aware architecture for personalized elderly care in smart environments. In: 1st EAI International Conference on Smart Grid Inspired Future Technologies (2016)

Optimizing Sliding Performance in iOS

Qin Zhao[1,2], Qi Qi[1,2(✉)], Lejian Zhang[1,2], and Qiwei Shen[1,2]

[1] State Key Laboratory of Networking and Switching Technology,
Beijing University of Posts and Telecommunications, Beijing 100876,
People's Republic of China
zhaoqin192@gmail.com,
{qiqi, zhanglejian, shenqiwei}@ebupt.com
[2] EBUPT Information Technology Co., Ltd, Beijing 100191,
People's Republic of China

Abstract. How to improve iOS sliding performance has always been the focus of iOS application optimization. This paper analyzes the principle of AutoLayout and Frame view layout, the opportunity of network loading, CPU and GPU performance consumption during sliding process. First, we provide the appropriate solution to avoid using AutoLayout, and adjust the time of network loading by preloading to reduce the waiting time dynamically. Pre-cache and asynchronous rendering to reduce the main thread CPU consumption is implemented to reduce the main thread CPU consumption, and at the same time, GPU consumption is reduced by asynchronous rendering. Finally, verify the feasibility and effectiveness of the optimization scheme by experiments. It is verified that the percentage of the main thread CPU consumption decreases by 17.2% and FPS increases from 37 Hz to 60 Hz.

Keywords: Sliding performance · AutoLayout · Pre-cache
Asynchronous · FPS

1 Introduction

The operating systems of iOS provide many UI views for users to browse more information via sliding up and down. In fact, developers can not request all resource from the Internet. In this scenario, operating systems need to load latest data after exploiting exhaustive search. Regardless of iOS or Android, network action is expensive because it costs much resource such as time, network traffic, electricity and so on. But the network situation of mobile phones is so bad in some cases that users have to wait until application receives network response which causes terrible user experience. On the other hand, after obtaining the resource from remote server, device should visualize data on the hardware screen. The CPU and GPU would finish work respectively to supply cache data for rendering on screen. The heavy load caused by CPU and GPU could lead to frame loss and set a delay response after users touching screen. There are many factors which affect the performance of CPU or GPU. AutoLayout based on Cassowary makes iOS layout simple and quick [1], while the improper way using AutoLayout or high load operation would drag the CPU. Blended layers, misaligned images and off-screen rendering are the killer of GPU, where off-screen rendering affects dramatically because it wastes a lot of performance to rendering off-screen images. That is to say, the tardiness of network and visualize data would affect sliding performance in iOS and UE (user experience). It is important that how to discover these key points and solve the knotty problems.

This paper is committed to optimizing sliding performance to improve user experience. From discovering the issues of AutoLayout and off-screen rendering performance, this paper proposes some method, such as pre-cache, asynchronous rendering to reduce the heavy burdens of device. So that many layout and rendering sites can be resolved, to improve the sliding performance of application. Due to the similar the hardware system or foundation framework for iOS or Android platforms [2], some optimization methods mentioned in this paper can be applied in other platforms.

2 Performance Optimization

The optimizing performance of application focuses on code structure optimizing and operational performance optimizing. References of Method for Mobile User Interface Design Patterns Creation for iOS Platform [3] gives guidelines for developers work in a high level of usability quality purpose, which belongs to structure optimizing. While it cannot promote performance when application is running. And References of On-device Objective-C Application Optimization Framework for High Performance Mobile Processors proposes a methodology to tailor a given Objective-C application and its associated device-specific shared library codebase using on-device post-compilation code optimization and transformation [4] that modified runtime library of iOS to acquire a high performance. But it would lose many features of runtime library, for example, the magic feature of using JavaScriptCore and runtime replace some method when calling some object message. Maybe there are other approach optimizing performance, and developer can avoid incorrect way that drag CPU and GPU of mobile device.

2.1 Preloading

With the development of mobile communication and the wide coverage of WIFI, mobile devices have better network services. However, network request is still an expensive operation. On the one hand, it costs a good deal of traffic, and developers need to take full advantage of the returned data as much as possible. On the other hand, network response time is unpredictable, which may cause bad user experience because of long waiting time. These two points are more obvious in iOS sliding. Initialization request data should not be too much, because users will not browse related information and waste a lot of traffic. So application should load the data according to the number of pages. Usually, it is time to load when slide to the bottom of the list page with showing a load animation and making network request. The drawback is that sliding page will stop until the network responds to new data, which wastes time for users. As a result, our primary goal is preloading, that is processing network request before reaching the bottom of the list. In this way, application has obtained new data before sliding to the bottom for display. Through predicting the users' behavior, it could effectively save the traffic, and make user use application without waiting for network requests. Developers can change the network load time despite they could not determine the network situation. In another word, the preloading method is not optimizing network request but optimizing the opportunity for network requests.

After demonstrating the correctness of preloading, the timing of preloading should be taken into account. Normally, setting the fixed threshold is a simply approach. For example, we set 0.7 as a threshold, and it will process network request when sliding at the 70% of the total height. The corresponding code is as follows (Fig. 1):

```
// threshold
CGFloat threshold = 0.7;
// currentPage
int currentPage = 0;
// slide up logic
- (void)scrollViewDidScroll:(UIScrollView *) scrollView {
    // current slide offset
    CGFloat offsetHeight = scrollView.contentOffset.y + scrollView.frame.size.height;
    // current total height
    CGFloat totalHeight = scrollView.contentSize.height;
    // the ratio of current height to total height
    CGFloat ratio = offsetHeight / totalHeight;
    // exceed the threshold and make request
    if (ratio >= threshold) {
        currentPage += 1;
        // request next page data
    }
}
```

Fig. 1. Fixed threshold

The code is very simple, but in fact it would not avoid the following problem: As the number of pages increasing, the height of the list will continue growing. A fixed threshold will lead to the growth of the height of unviewed page. In order to compare the waste of network resources, we assume that each table cell has the same height. Therefor the size of the data is able to reflect the height of the view. The threshold is shown in Table 1.

Table 1. The effect of fixed threshold.

Page	TotalNum	TimeNum	Diff
1	10	7	3
2	20	14	6
3	30	21	9
4	40	28	12
5	50	35	15
6	60	42	18

Page represents the number of pages and TotalNum represents the total count of data respectively. TimeNum indicates the data which has been viewed when preloading. Diff shows the amount of data which has not been browsed. It is displayed that the amount of data which has not yet been browsed will increase as the number of pages growing. The opportunity for preloading is kept in advance and leads to a lot of data being loaded that users would not browse. Furthermore, it causes the waste of application traffic.

To above issues, we design a new method to optimize the preloading: For each page, it is specified that 70% of the amount of new data is set as the threshold. If the sliding height exceeds this threshold, application will request new data and change the threshold again. The code is depicted in Fig. 2.

```
//preloaded when reach to 70% of the last page
CGFloat threshold = 0.7;
// current page
int currentPage = 1;
// the amount of data per page
int perPageNum = 10;
// slide loop logic
- (void)scrollViewDidScroll:(UIScrollView *) scrollView {
    // current slide offset
    CGFloat offsetHeight = scrollView.contentOffset.y + scrollView.frame.size.height;
    // current total height
    CGFloat totalHeight = scrollView.contentSize.height;
    CGFloat ratio = offsetHeight / totalHeight;

    // the amount that need to make a network request
    CGFloat needRead = itemPerPage * threshold + currentPage * itemPerPage;
    CGFloat totalItem = itemPerPage * (currentPage + 1);
    // dynamically adjust the threshold
    CGFloat newThreshold = needRead / totalItem
    // exceed the threshold and make request
    if ratio >= newThreshold {
        currentPage += 1;
        // request next page data
    }
}
```

Fig. 2. Dynamical threshold

With the growth of page amount, through adjusting the threshold, the amount of data which is not viewed remains within a stable range, and as a result network resources could be saved. Shown in Fig. 3, as the number of pages increasing, the threshold will dynamically grow to delay the preloading time.

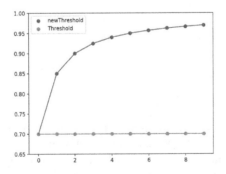

Fig. 3. The curve of dynamically threshold and fixed threshold

2.2   The Bottleneck of AutoLayout Performance

AutoLayout [7] is the implementation of the UI layout program after iOS6, which can easily solve the UI adaptation problem. Nevertheless, we should abandon this technical program to get higher sliding performance. It is because that compared with the traditional Frame layout, AutoLayout makes the design of UI convenient, but it would affect CPU performance when running application. The traditional Frame layout is to specify the location of a UI view in the parent view, which must include the coordinates of the axis x, y and the length and width of the view itself. AutoLayout is based on the Cassowary algorithm that adds a lot of constraints to the view, such as the distance to the left and top of the parent view and so on. All of these constraints are abstracted as a set of linear equations or inequalities. Finally, the operating system get the x, y coordinates and the width and height by solving the set of linear equations or inequalities. However, the calculation of linear equations requires CPU to consume. If there are a large number of views using AutoLayout, it will need CPU to solve multiple sets of linear equations at the same time. The refresh rate of iOS is 60 Hz, which is vital for optimizing sliding performance. If the CPU solves the layout for more than 16.67 ms (1/60 s = 16.67 ms), it will cause the data of this frame not to be prepared in the buffer. When V-sync signal coming, it is inevitable that FPS [8] will be declined to impress sliding performance if the required rendering data is not available when the buffer data is read.

Figure 4 is obtained by randomly generating N views on iOS10.2.1 iPhone7 and the layout modes are AutoLayout and Frame respectively. The abscissa in the graph represents the number of render views, and the ordinate represents the time to render the views.

From Fig. 2, Frame mode has better performance in all cases. For example, it takes about 11.7 ms to render when the number of views is 100, and AutoLayout needs 32.0 ms at the same circumstance. The rising speed of the curve of AutoLayout is also significantly greater than that of Frame. And according to iOS rendering frequency, we can see that when the rendering time is greater than 16.67 ms, it will certainly generate block. As seen from Fig. 3, it will produce the performance problem when generating

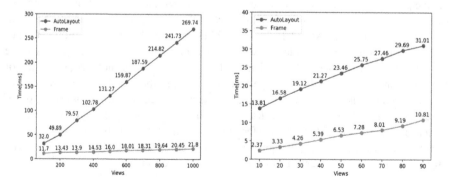

Fig. 4. AutoLayout vs. frame

approximately 20 views on the experimental device under AutoLayout mode. Relatively, Frame layout only consumes about 16 ms to generated about 500 views.

Therefore, if it needs to maintain a high FPS or high performance during the sliding process, we should not select AutoLayout as technical proposal. While Frame layout is more cumbersome, it can be compensated by efficient operation to provide better sliding performance. Obliviously the first step in optimizing sliding performance is to Frame layout.

2.3   Pre-cache

The widgets of UI Views (such as UITableView [9], UICollectionView [10]) in iOS programming exist reuse mechanism [5], which only stores the current cell displayed on the screen, regardless of all the list cell. The biggest advantage of this mechanism is to save memory space. Assuming that every data generates a cell, the memory will soon be exhausted if there are a large amount of data. However, there is a flaw in the reuse mechanism that different data have different styles. The application needs to recalculate the layout information in real time and then display it. In the process of rapid sliding, a large number of calculation has a bad impact on the performance of the main thread CPU. How to ease the pressure of the CPU during the sliding process is very important for optimizing performance. This paper employs pre-cache to solve the problem. Pre-cache is to calculate the view layout in advance and cache them. Pre-cache creates asynchronous thread to parse the layout model after getting data from network, and each model stores the information for displaying. The layout of the Cell is uniquely determined by the contents of the model, so application will calculate the layout information in advance based on the content of the model and store these data.

It will notify the main thread to refresh the UI after all data has been analyzed. Although it is necessary to continually update the UI during the rapid sliding of the list, pre-cache can reduce the operating load of the CPU during sliding because it can read UI view data directly without recalculating.

2.4   Asynchronous Drawing

Via the pre-cache processing, it has been reduced the burden of CPU during sliding to a great degree. However, we find that there are still some points to optimize after analyzing the performance loss of CPU during sliding process. The function drawInText of UILabel is responsible for rendering text that is running on the main thread. It will be not serious if the number of UILabel [11] is small, and has not become a major constraint on the impact of sliding performance. If UILabel needs to display a lot of text, it would make CPU performance degradation because it occupies a lot of resources of the main thread. We use self-defined CALayer [12] layer to ease the burden of main thread CPU by transferring the timing of rendering to asynchronous thread, using CoreText [13] Framework and asynchronous thread to draw text. The UILabel and AsyncLabel architecture of iOS is depicted in Fig. 5.

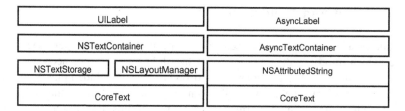

Fig. 5. UILabel architecture diagram vs AsyncLabel architecture diagram

Both AsyncLabel and UILabel are based on CoreText. And CoreText is also the basic framework for all text and image widget in iOS programming. AsyncLabel keeps the basic information of the NSAttributedString, and the NSTextStorage and NSLay-outManager layer are greatly simplified. The biggest difference between UILabel and AsyncTextContainer is the rendering time of AsyncTextContainer layer occurs on asynchronous threads, which finally renders the picture in the main thread that described by text information through the CTFrame [14], CTLine [15], CTRun [16]. Obviously, asynchronous drawing can reduce the CPU load during the sliding process further.

2.5   Asynchronous Rendering

After pre-caching, we can find that the main thread of CPU usage is relieved, but it is still not smooth during sliding. Then we turn our attention to easing the burden on GPU. Detecting FPS by the Instruments, a fantastic tool for monitoring all performance of iOS device, we find that the biggest factor which affects GPU performance is off-screen rendering. Off-screen rendering composites a part of the layer tree into a new buffer (which is off-screen, i.e. not on the screen), and then that buffer is rendered onto the screen. Generally, we do our best to avoid off-screen rendering, because it costs too much.

In the client of iOS, the business scene which can trigger off-screen rendering is to set the users' pictures as rounded corners. How to set the image circular efficiently is one of the key factors to improve the sliding performance. The traditional way to set the circular is to cover the CALayer, but it would cause off-screen rendering and consume GPU performance. Of course, if the client can get the circular image from server directly, there is no GPU rendering problem. For the same picture, different business scenes need different shapes of pictures, for example, some places require a rectangle and some place require rounded corners. So the fundamental solution is to instruct the client to handle rounded pictures locally and the essence of the problem translates into how to set the rounded image efficiently. The optimization scheme designed in this paper is using asynchronous rendering. We import the original image resource to an asynchronous thread and then use the underlying CoreGraphics in the asynchronous thread for rounded corners or other effects. The processing here does not create rounded corners by setting the CALayer, but it employs the Bezier curve to cut the original picture as a new picture resource. Finally, the processed image is passed to the main thread to display. In this case, it could improve the FPS through asynchronous rendering to a large extent.

3 Experiment and Validation

In the optimization function discussed above, the preload can prepare the data in advance, which improves the smoothness of the sliding and do not need to wait for network to respond [6]. The previous comparison has been concluded that AutoLayout affects sliding performance dramatically. Ignoring the two factors that have been identified to affect the sliding performance, we set two groups of layout to compare the optimization effect in the pre-cache, asynchronous rendering, asynchronous loading. The first set of data is not set the above optimization point, and the second set of data is on the contrary. Test environment is macOS 10.12.2 system, XCode 8.2.1, iPhone 7, iOS 10.2.1 system.

It can be seen from the above experimental data that in about 30 s of time-consuming, depicted in Figs. 6 and 7. The percentage of the main thread CPU consumption decreases by 17.2%, and at the same time, asynchronous thread CPU time-consuming increases by 19.4%. The results are consistent with the expectations

Weight		Self Weight	Symbol Name
22.94 s	73.9%	0 s	Main Thread 0x6a22e
1.07 s	3.4%	0 s	_dispatch_worker_thread3 0x6a265
627.00 ms	2.0%	0 s	_dispatch_worker_thread3 0x6a23d
532.00 ms	1.7%	0 s	_dispatch_worker_thread3 0x6a26e
517.00 ms	1.6%	0 s	_dispatch_worker_thread3 0x6a249
420.00 ms	1.3%	0 s	_dispatch_worker_thread3 0x6a2b0
365.00 ms	1.1%	0 s	_dispatch_worker_thread3 0x6a2c6
353.00 ms	1.1%	0 s	_dispatch_worker_thread3 0x6a23f
259.00 ms	0.8%	0 s	_dispatch_worker_thread3 0x6a2c0
151.00 ms	0.4%	0 s	_dispatch_worker_thread3 0x6a27a
150.00 ms	0.4%	0 s	_dispatch_worker_thread3 0x6a2d6
120.00 ms	0.3%	0 s	_dispatch_worker_thread3 0x6a2e5
50.00 ms	0.1%	0 s	_dispatch_worker_thread3 0x6a2e9
40.00 ms	0.1%	0 s	_dispatch_worker_thread3 0x6a262
35.00 ms	0.1%	0 s	_dispatch_worker_thread3 0x6a2d9
17.00 ms	0.0%	0 s	_dispatch_worker_thread3 0x6a264
17.00 ms	0.0%	0 s	_dispatch_worker_thread3 0x6a26b
16.00 ms	0.0%	0 s	_dispatch_worker_thread3 0x6a23e
7.00 ms	0.0%	0 s	std::__1::__thread_proxy<std::__1::tuple<com::bytedance::push::WebSocketClient::WebSocketClient 0x6a269
6.00 ms	0.0%	0 s	WebCore::ThreadGlobalData::~ThreadGlobalData 0x6a27e
1.00 ms	0.0%	0 s	WebCore::StorageThread::threadEntryPoint 0x6a289
1.00 ms	0.0%	0 s	asio::detail::posix_thread::func<asio::detail::resolver_service_base::work_io_service_runner>::run 0x6a26a

Fig. 6. The analysis of unoptimized application CPU consumption

Weight		Self Weight	Symbol Name
17.31 s	56.7%	0 s	Main Thread 0x72d23
1.72 s	5.6%	0 s	_dispatch_worker_thread3 0x72d48
1.61 s	5.2%	0 s	_dispatch_worker_thread3 0x72db4
1.51 s	4.9%	0 s	_dispatch_worker_thread3 0x72d71
1.31 s	4.2%	0 s	_dispatch_kevent_worker_thread 0x72d69
1.01 s	3.3%	0 s	_dispatch_worker_thread3 0x72ded
716.00 ms	2.3%	0 s	_dispatch_worker_thread3 0x72d84
696.00 ms	2.2%	0 s	_dispatch_worker_thread3 0x72d72
578.00 ms	1.8%	0 s	_dispatch_worker_thread3 0x72dee
536.00 ms	1.7%	0 s	_dispatch_worker_thread3 0x72ddd
316.00 ms	1.0%	0 s	_dispatch_worker_thread3 0x72d70
271.00 ms	0.8%	0 s	_dispatch_worker_thread3 0x72d3f
115.00 ms	0.3%	0 s	_dispatch_worker_thread3 0x72e00
110.00 ms	0.3%	0 s	_dispatch_worker_thread3 0x72d40
88.00 ms	0.2%	0 s	_dispatch_worker_thread3 0x72df3
18.00 ms	0.0%	0 s	_dispatch_kevent_worker_thread 0x72d3d
11.00 ms	0.0%	0 s	_dispatch_worker_thread3 0x72d5d
9.00 ms	0.0%	0 s	std::__1::__thread_proxy<std::__1::tuple<com::bytedance::push::WebSocketClient::WebSocketClient 0x72d6b
6.00 ms	0.0%	0 s	RunWebThread 0x72d7e
4.00 ms	0.0%	0 s	_dispatch_worker_thread3 0x72d67
1.00 ms	0.0%	0 s	asio::detail::posix_thread::func<asio::detail::resolver_service_base::work_io_service_runner>::run 0x72d6c
1.00 ms	0.0%	0 s	-[NSRunLoop run] 0x72d7f

Fig. 7. The analysis of optimized application CPU consumption

after optimization. And it is confirmed that the increasement in asynchronous thread consumption is slightly greater than the reduced consumption in main thread because the growth in the number of threads on the CPU also has a certain impact. Via pre-caching and asynchronous rendering, the main thread of the work can move to the asynchronous thread which reduces the main thread consumption.

On the other hand, we also compare the GPU performance before and after the asynchronous rendering (from the perspective of FPS), shown in Figs. 8 and 9. Asynchronous rendering mainly improves the performance of the GPU. Before optimizing, FPS is around 37 Hz during fast sliding process, and after optimization, FPS increases to 60 Hz. In the process of rapid sliding, it is very smooth in line with the expected optimization.

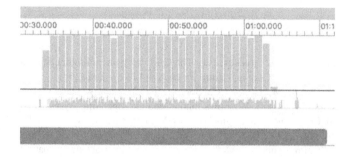

Fig. 8. Unoptimized application of FPS data with GPU

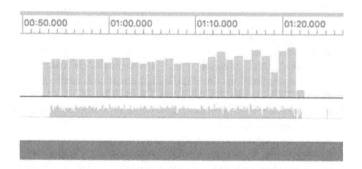

Fig. 9. Optimized application of FPS data with GPU

4 Conclusions

Sliding performance optimization in iOS is the key to providing a good user experience, especially in the new application [18]. In this paper, we constantly adjust the timing of requesting network data through dynamic preloading and get data without user awareness to save network resources, as well as reducing the sliding process to wait for the network response time. And then from the performance of the CPU and GPU, we use Frame layout, pre-cache, asynchronous rendering to reduce CPU performance loss, and use asynchronous rendering to avoid GPU loss brought by off-screen rendering. It is found that all of these measures make the usage rate of CPU declining 30%, the usage rate of GPU declining 40%, FPS rise from 37 Hz to 60 Hz, and achieve a very smooth sliding effect. In the future, the iOS performance can be optimized by the popular machine learning mechanism [18–20].

Acknowledgement. This work was supported in part by the (1) National Natural Science Foundation of China (No. 61671079, 61771068, 61471063) (2) Beijing Municipal Natural Science Foundation (No. 4182041).

References

1. Badros, G.J., Borning, A., Stuckey, P.J.: The Cassowary linear arithmetic constraint solving algorithm. ACM Trans. Comput.-Hum. Interact. **8**(4), 267–306 (2001)
2. Novac, O.C., Novac, M., Gordan, C., Berczes, T.: Comparative study of Google Android, Apple iOS and Microsoft Windows phone mobile operating systems. In: 2017 14th International Conference on Engineering of Modern Electric Systems (EMES). Oradea, Romania, pp. 154–159 (2017)
3. Wetchakorn, T., Prompoon, N.: Method for mobile user interface design patterns creation for iOS platform. In: 2015 12th International Joint Conference on Computer Science and Software Engineering (JCSSE), Songkhla, Thailand, pp. 150–155 (2015)
4. Bournoutian, G., Orailoglu, A.: On-device Objective-C application optimization framework for high performance mobile processors. In: Design, Automation & Test in Europe Conference & Exhibition (DATE), Dresden, Germany, pp. 1–6 (2014)
5. Ferreira, P.: Reclaiming storage in an object oriented platform supporting extended C++ and Objective-C applications. In: Proceedings 1991 International Workshop on Object Orientation in Operating Systems, Palo Alto, CA, USA, pp. 100–102 (1991)
6. Gutierrez, A., Dreslinski, R.G., Wenisch, T.F.: Full-system analysis and characterization of interactive smartphone applications. In: 2011 IEEE International Symposium on Workload Characterization (IISWC), Austin, TX, USA, pp. 81–90 (2011)
7. Develop Apple. https://developer.apple.com/library/content/documentation/UserExperience/Conceptual/AutolayoutPG/index.html
8. WikiPedia. https://en.wikipedia.org/wiki/FPS
9. Develop Apple. https://developer.apple.com/documentation/uikit/uitableview
10. Develop Apple. https://developer.apple.com/documentation/uikit/uicollectionview
11. Develop Apple. https://developer.apple.com/documentation/uikit/uilabel
12. Develop Apple. https://developer.apple.com/reference/quartzcore/calayer
13. Develop Apple. https://developer.apple.com/documentation/coretext
14. Develop Apple. https://developer.apple.com/documentation/coretext/ctframe
15. Develop Apple. https://developer.apple.com/documentation/coretext/ctline
16. Develop Apple. https://developer.apple.com/documentation/coretext/ctrun-61n
17. Xu, P., Yin, Q., Huang, Y., Song, Y.-Z., Ma, Z., Wang, L., Xiang, T., Kleijn, W.B., Guo, J.: Cross-modal subspace learning for fine-grained sketch-based image retrieval. Neurocomputing **278**, 75–86 (2018)

18. Ma, Z., Xue, J.-H., Leijon, A., Tan, Z.-H., Yang, Z., Guo, J.: Decorrelation of neutral vector variables: theory and applications. IEEE Trans. Neural Netw. Learn. Syst. **29**(1), 129–143 (2018)
19. Liu, W., Cao, J., Yang, L., Xu, L., Qiu, X., Li, J.: AppBooster: boosting the performance of interactive mobile applications with computation offloading and parameter tuning. IEEE Trans. Parallel Distrib. Syst. **28**(6), 1593–1606 (2017)
20. Ma, Z., Rana, P.K., Taghia, J., Flierl, M., Leijon, A.: Bayesian estimation of Dirichlet mixture model with variational inference. Pattern Recogn. **47**(9), 3143–3157 (2014)

A Dialog Robot Based on WeChat

Xiaoyi Chen, Jing Wang, Qiwei Shen, Qi Qi$^{(\boxtimes)}$, and Jingyu Wang

State Key Laboratory of Networking and Switching Technology,
Beijing University of Posts and Telecommunications,
Beijing 100876, People's Republic of China
{chenxiaoyi,wangjing,shenqiwei,qiqi,
wangjingyu}@ebupt.com

Abstract. WeChat is one of the most popular instant messaging applications in the world. It has now become an important access to variety business systems for billions of users. The vast majority of companies want to provide their business services onto WeChat in order to gain advantage in fierce market competitions. However, as far as we know, today it is not easy to access WeChat with business service. In this paper, we propose a framework to integrate business services and WeChat. On the basis of this framework, companies or entrepreneurs can provide their business services on WeChat easily. Finally, we use a case study to demonstrate how our service can be used in helping tickets sells and statistical analysis.

Keywords: WeChat on Web · Dialog robot · SNSs

1 Introduction

In recent years, since the Internet has become more developed, many innovative information services have been created. Of which, Social Networking Services (SNSs) have been well-received by the public. Using the Internet to maintain interpersonal relationships through SNSs has become a part of modern life. SNSs has unique capability to build mobile social networks among people who has the same interests. As network information technology has increasingly advanced, there is a growing trend of people using electronic devices. With the rise of social network services (SNSs), people are using SNSs more frequently, SNSs have gradually replaced many traditional methods of contacting, such as sending emails, typing text messages, or chatting on the phone. There are lots of famous SNSs providers include WeChat [1], KakaoTalk [2], and WhatsApp [3] etc. Because of its capacity to connect people, more and more industry companies devoted a lot of manpower and resources to develop business service among Social Network Services.

However, as far as we know, it is not easy to let industry companies provide their business services for SNSs providers. Because it requires lots of efforts to satisfy the frequently updated APIs offered by SNSs providers. Like WeChat often updates their APIs to provide more comprehensive services.

WeChat is one of the most popular instant messaging applications in the world. It provides text, image, voice and video communication service for smart terminal users.

© ICST Institute for Computer Sciences, Social Informatics and Telecommunications Engineering 2018
P. H. J. Chong et al. (Eds.): SmartGIFT 2018, LNICST 245, pp. 122–132, 2018.
https://doi.org/10.1007/978-3-319-94965-9_13

Besides the basic functionality of sending message. They can also use multiple convenient services such as WeChat Moments and Official Accounts to share and publish information. Due to the free and abundant services, it not only changed the way people communicate, but also triggered a new media revolution, become the most popular new media communication tools. At the end of 2016, it has covered more than 90% smart phones and the monthly active users reached 889 million from over 200 countries [4].

In February 20, 2014, Tencent announced the launch of QQ browser for WeChat version, that is, WeChat on Web [5]. The purpose of WeChat on Web is to bring more convenient way for users to communicate. It gets through the WeChat mobile version and web version. After that, users can directly send, receive, and even transfer between the computer and mobile phone files in the web browser. It uses its own WeChat Web API to communicate with WeChat server, and it can achieve most of the functions on the WeChat mobile version. The detailed introductions of this API will be given in Sect. 3. Compared to the WeChat Public Platform [6], it has the following three advantages:

(1) It provides more basic functions, which can be extended easily.
(2) If developers who want to use WeChat public API to provide services, they should register a WeChat Official Account first. As we know, the registration process of WeChat Official Account is very cumbersome. It provides few functionalities if you pay more money to upgrade the account. It would be a laborious and costly thing for users who use our framework.
(3) As we use WeChat Web API, developers who use our framework can provide personalized services use any WeChat account (business account or personal account). It can also provide services in WeChat group, which is the Official Account can not do. Therefore, we use WeChat Web API instead of WeChat public API to achieve our framework.

However, we also need to face these challenges: (1) As WeChat do not provide official documentation for Web APIs and it may evolve over time, developing business services over WeChat requires lots of development efforts; (2) As our framework can provide unified API to the developers, it is difficult to abstract all kinds of variety WeChat messages into a standard interface; (3) As the core of our framework is dialogue robot, how to identify the topic and keywords in the conversation is the key to provide good service to users.

To address these three challenges, in this paper, we plan to develop a service-based framework for mobile dialog service using WeChat. Using this framework, developers will save lots of time in reading API specifications of WeChat Public Platform or coding system, since they can easily access their business services through our framework. This framework can also provide other services such as knowledge integration, dialogue analysis and statistic.

2 Related Work

Yitong et al. [7] proposes a framework to integrate WoT and WeChat platform. They designed and realized the system basing on WoT and WeChat public API. Xiang et al. [8] discussed the features of WeChat public platform, and then proposed a smart university campus information dissemination framework based on WeChat public platform. Maohong et al. [9] provided mobile learning resources based on WeChat public platform, and applies WeChat in mobile learning, improve the flexibility of learning, build more good autonomous learning and collaborative learning environment, and promote the learning effect.

Mei et al. [10] propose a framework for providing mobile dialog services using WeChat. They provided a framework with which industry companies can easily provide their IT services using WeChat. This framework can serve as the bridge between users and various information and business functionalities. Their previous study [11] introduced a framework on the social messaging integration in PaaS. They introduced how to provide mobile dialog service in details, which can not only be deployed on PaaS, but also on other standalone VMs.

The above services or frameworks are all based on WeChat public platform. Considering we use WeChat web API instead of WeChat public API, developers can configure their business services use their own WeChat account, rather than spending lots of time in reading API specifications of WeChat Public Platform, design, coding, and debugging or registering the WeChat Official Account. As we all know, the registration process of WeChat Official Account is very cumbersome.

2.1 Technical Architecture

Figure 1 shows the technical architecture of our framework. Our framework includes four major parts: (1) Web platform (2) WeChat on Web (3) Rest [12] API and (4) Our engine for WeChat Integration.

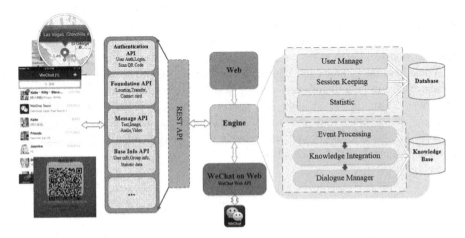

Fig. 1. System architecture for mobile dialog service framework using WeChat.

Web platform provides a simple configuration interface of dialog robot and a variety of statistical graph for the users who are not have the ability to develop the application but still want to provide dialog robot server using their own WeChat account.

The WeChat on Web layer are provided by WeChat Web API, the detailed introductions of WeChat Web API are already given in next section.

The role of Rest API layer is to abstract the business logic into a few main interfaces to the developers or company, it majorly consisted of following four parts: the authentication API, the foundation API, the message API and the base info API as well. These four parts are separated independently in order to make it easy for developers to use it but worked interconnected. The authentication API is used to check the user's legitimacy. It can get the login QR code of WeChat on Web and give interface to check if the QR code is scanned or if the user is logged in. The purpose of the message API is to provide users a message handler process used to receive and send the message from or to the WeChat, which is the fundamental function of instant messaging. Meanwhile, foundation API provide basic functions of WeChat like send location, contact card and transfer with friends. It waits for the event and send the key value to the developer as a response. The forth part is base info part, base info API is for establishing, querying and deleting user or group info.

The engine is the main part of our system, the role of the engine includes the following aspects:

(1) User/Robot manage. In our framework, any WeChat account can be used as a user, but also as a robot. A user can manage multiple robots; a robot can only be supervised by a user. Users can manage each robot in our web platform. All information of users and robots is managed by this module.

(2) Session keeping. As we provide dialog robot using WeChat, each robot need to keep connecting with WeChat use WeChat Web API. In this module we use the Actor model. The actor model in computer science is a mathematical model of concurrent computation which treats actors as the universal primitives of concurrent computation. In response to a message that it receives, an actor can: make local decisions, create more actors, send more messages, and determine how to respond to the next message received. The Actor model can be used as a framework for modeling, understanding, and reasoning about a wide range of concurrent systems. In our framework, each robot is supervised by an Actor. Each Actor has its own context, it manages and maintains all the information of the session, if a robot is disconnected with WeChat, it will send an event to inform the robot's supervisor immediately.

(3) Statistic. The responsibility of the statistic module is to analyze all the conversational data statistically and to present the statistical results to the Web platform for the user's query.

4) Dialogue engine. The dialog engine consists of three parts: Event Processing, Dialogue Manager and Knowledge Integration. In this framework, all messages associated with WeChat are abstracted as events. Events consists of type and content. Event Processing is to abstract the message and do the appropriate treatment according to the different types of events. For example, text messages

will be send to Knowledge Integration for semantic analysis. The role of Dialogue Manage is to maintain the context of the session. Because a robot may serve multiple clients at the same time. In addition, the same sentence in different contexts will be expressed as different semantics. So, in our framework, we use a simple state machine model and keyword matching to handle this situation. We use state to represent different context, the same text in different states will be treated differently. In the same context, what the servers need to do is matching the keywords which are configured by users or developers and feedback information. For example, if the user send "I want to buy a train ticket from Beijing to Shanghai tomorrow", the words "train ticket", "Beijing", "Shanghai" and "tomorrow" would be matched, and the system would get the ticket information from Beijing to Shanghai tomorrow and send back a detail train list to the user. That is what Knowledge Integration do. Its purpose is to find the most approximate response. Its will first find candidates response from database. If there are multiple candidates present. It will choose the highest priority result according to the different semantics. If no candidate response exists, it will send the message to the Turing robot (Third part knowledge base) to find the most appropriate answer.

2.2   WeChat Web API

WeChat have its own customized protocol for Web. Its protocol and data structure have some different between WeChat on iPhone [13] or Android [14, 15]. Figure 2 shows the flow chat of WeChat on Web. Each of the above boxes is an interface, each interface communicates with each other through data, and all of these APIs form the whole WeChat Web framework. It works like this:

(a) The client first gets an UUID from WeChat server, UUID can be understood as a token, it is used for generate QR code.
(b) Then the client sends the UUID to server and get the QR code as response.
(c) The client initiates a request to the server repeatedly to query whether the user has already scanned the code.
(d) If the user scans the QR code, server will return a series of verify message, it includes user identity, session message and token.
(e) The client can use these verify message to fetch detail information of users from server, this information includes user basic information, friend contact and group contact etc.
(f) The client informs the server that the message has been received and establishes a long connection with the server, and then all messages that interact with the server are sent on this link.
(g) If there are new message received, the server will inform the client immediately. This message can be a new friend invitation, a contact card from friend or a red packet from group. Each message has its own specific format. It mainly includes the sender of the message, the recipient of the message, the message type and the content.
(h) Meanwhile, the client can also send various types of messages proactively to the server by its message API.

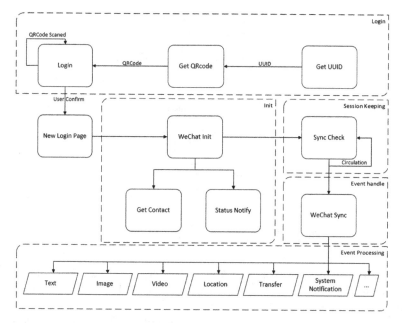

Fig. 2. Flow chart of WeChat on Web.

With this interface, our framework can be achieved with all aspects of WeChat interaction. It can support, including messaging, friend management, group management, personal information settings and all the features of the WeChat on Web, which is the basic of providing dialog robot for WeChat.

2.3 Text Classification

In order to allow robots to provide more humane services, we must first understand the user's input. The most common practice is to analyze the text entered by the user. This process can be divided into two parts, one is the text segmentation, the other is the clustering of text. As we all know, word is the minimum meaningful unit of languages. However, unlike English and other western languages, there is no natural delimiter between Chinese words and even no uniform smallest semantic units.

Many standard machine learning techniques have been applied to automated text classification problems, and kNearest Neighbor algorithm (kNN) and Support Vector Machine (SVM) have been reported as the top performing methods for English text classification [16, 17]. However, the studies on Chinese text classification are less sufficient compared with English and Chinese text has its own characteristic. As there is no natural delimiter between Chinese words, this means that the Chinese segmentation is necessary before any other preprocessing.

Luo et al. [18] give an efficient and effective approach to improve the performance of Chinese text classification. They study on Chinese text classification using character-based approach (N-gram) and word-based approach and propose the use of uni-gram, bi-gram and word features of length greater than or equal to three. They also

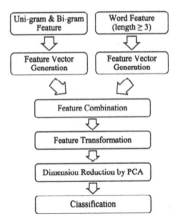

Fig. 3. Steps for Chinese text classification [17].

investigate a serial approach based on feature transformation and dimension reduction techniques to improve the performance. Figure 3 shows their steps for Chinese text classification based on the proposed approach.

After a lot of testing, we use proposed approach of *1 + 2-gram* feature set of *RBF* kernel with the weight coefficient α is assigned a value of 10. In our system, most of the dialogue are short dialogue, like "What's the weather today?". In this case, the algorithm offered by Luo and their team have shown a very good performance in the context of Chinese and English, which can fully meet the needs of robot dialogue analysis in our system.

2.4 Keywords Matching

In our system, it is necessary to determine whether a keyword appears in the text of the conversation. To determine whether a word appears in the text, the commonly method take the pattern string to match the text string word by word, in which case the time complexity is $O(P * T)$ (P represent the length of pattern string, T represent the length of text string), the improved algorithm represented by KMP algorithm, the time complexity is $O(P + T)$, which has a highly matching efficiency. However, in our system, it is often necessary to take multiple pattern strings to match, in which case the time to match a text becomes $O(L * (P + T))$ (L represent the number of pattern string). And usually L is greater than 100. At this case the matching speed is far less than our expectations.

Based on our business scenario (the vast majority of the text is short dialogue), we come up with a method using text string to match the pattern string. First, the pattern string is pre-processed with the form of the K-V stored in the Hashset table. We know that the time complexity of finding an element in a HASHSET is $O(1)$, so the total time complexity reduced to $O(K * 1)$ (K represent the number of segments divided by text string, and the time complexity of searching for a string in a pattern string is constant). And K is less than 10 in most cases. So the speed of matching in this way is very fast, which greatly enhance the efficiency of our robot's text analysis.

3   Case Study

In this section, we present a case study on how our framework can be used by organizing a ticket system. In this case study, we use a train ticket booking for illustration. We demonstrate the capabilities from the following 4 aspects.

(1)   Ticket bookings. The robot can provide buying guidance for the users (Fig. 4). And users can simply pay the ticket by red packet. All the operations are user-friendly.

(2)   Ticket inquiries. Based on the rest APIs provided by the ticket company, robot can provide real-time ticket query.

(3)   Ticket reminder. Before the plane or train departs, the robot will inform the user of preparing in advance by proactively pushing the message.

(4)   Statistics. According to the robot's dialogue message, this information can be analyzed from different dimension, such as finding the most active period of users, analyzing the most common topic for users (Fig. 5), and doing some business-related statistics (e.g. Finding out ten highest sales days in one year (Fig. 6)). Meanwhile, these statistics will display on the Web page for developers to view in the form of charts or figure.

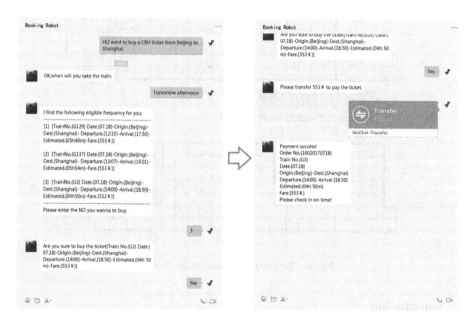

Fig. 4.   Payment guidance.

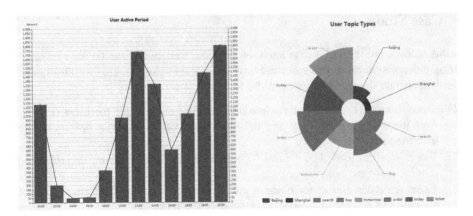

Fig. 5. Statistic data of user.

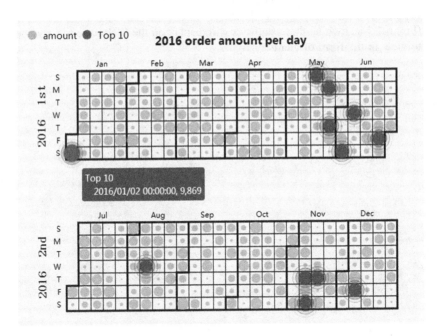

Fig. 6. Top 10 order quantity in 2016.

4   Conclusion

Providing business applications based on WeChat has gradually become popular in recent years. Many industry companies have provided their service on WeChat in order to gain an advantage in the fierce market competition. But as far as we know, to most of developers, it is not easy to build such a dialog robot based on WeChat because WeChat is updated occasionally and it need to spend lots of efforts combining dialog robot with WeChat server. In this paper, we proposed a framework for providing dialog robot based on WeChat. We explained how our framework work and use a study case to display how to use our framework to provide service. We hope it can help developers reduce the development efforts. But in this version, our robot can only deal with the text message and the position message, maybe the voice and the other information is represented automatically [19, 20] and will be added in the future.

Acknowledgement. This work was supported in part by the (1) National Natural Science Foundation of China (No. 61671079, 61771068, 61471063) (2) Beijing Municipal Natural Science Foundation (No. 4182041).

References

1. WeChat. http://www.wechat.com/
2. KakaoTalk. http://www.kakao.com/talk
3. WhatsApp. http://www.whatsapp.com/
4. WeChat User Research Report (2017). http://tech.qq.com/a/20170424/004297.htm
5. WeChat on Web. https://wx.qq.com/?lang=en_US
6. WeChat Public Platform. https://mp.weixin.qq.com/
7. Yitong, H., Xiaozheng, L., Bingpei, D., Qinyi, C.: Web-of-things framework for WeChat. In: 2013 IEEE International Conference on Green Computing and Communications and IEEE Internet of Things and IEEE Cyber, Physical and Social Computing, pp. 1496–1500. IEEE Press, Beijing (2013)
8. Xiang, Y., Chang, D., Chen, B.: A smart university campus information dissemination framework based on WeChat platform. In: Zhang, R., Zhang, Z., Liu, K., Zhang, J. (eds.) LISS 2013, pp. 927–932. Springer, Beijing (2015). https://doi.org/10.1007/978-3-642-40660-7_138
9. Maohong, Z., Hui, L., Xingzhi, Z., Li, Z., Xiaoli, Z.: Research of mobile learning mode based on WeChat public platform. In: First IEEE International Conference on Computer Communication and the Internet, pp. 489–492. IEEE Press, Wuhan (2016)
10. Lijun, M., Hao, C., Yabin, D., Qicheng, L., Shaochun, L.: Providing mobile dialog services using WeChat. In: 2016 IEEE International Conference on Mobile Services, pp. 135–141. IEEE Press, California (2016)
11. Mei, L., Chen, H., Li, S., et al.: A service-based framework for mobile social messaging in PaaS systems. In: IEEE International Conference on Web Services, pp. 751–754. IEEE Press, New York (2015)
12. Leonard, R., Sam, R.: RESTful Web Services, pp. 299–314. O'Reilly, Newton (2007)
13. Feng, G., Ying, Z.: Analysis of WeChat on IPhone. In: 2nd International Symposium on Computer, Communication, Control and Automation, pp. 278–281. Atlantis Press, Singapore (2013)
14. Songyang, W., Yong, Z., Xupeng, W., Xiong, X., Lin, D.: Forensic analysis of WeChat on Android smartphones. In: Digital Investigation, vol. 21, pp. 3–10. Elsevier, Seattle (2017)
15. Lijun, Z., Fei, Y., Qingbing, J.: The forensic analysis of WeChat message. In: 2016 Sixth International Conference on Instrumentation & Measurement, Computer, Communication and Control, pp. 500–503. IEEE Press, Harbin (2016)
16. Yang, Y., Liu, X.: A re-examination of text categorization methods. In: 22nd Annual International ACM SIGIR Conference on Research and Development in Information Retrieval (SIGIR 1999), Berkeley, pp. 42–49 (1999)

17. Ma, Z., Xie, J., Li, H., Sun, Q., Si, Z., Zhang, J., Guo, J.: The role of data analysis in the development of intelligent energy networks. IEEE Netw. **31**(5), 88–95 (2017)
18. Luo, X., Ohyama, W., et al.: Automatic Chinese text classification using character-based and word-based approach. In: 12th International Conference on Document Analysis and Recognition, pp. 329–333. IEEE Press, Washington DC (2013)
19. Ma, Z., Teschendorff, A.E., Leijon, A., Qiao, Y., Zhang, H., Guo, J.: Variational Bayesian matrix factorization for bounded support data. IEEE Trans. Pattern Anal. Mach. Intell. **37**(4), 876–889 (2015)
20. Ma, Z., Xue, J.-H., Leijon, A., Tan, Z.-H., Yang, Z., Guo, J.: Decorrelation of neutral vector variables: theory and applications. IEEE Trans. Neural Netw. Learn. Syst. **29**(1), 129–143 (2018)

NO-V2X: Non-orthogonal Multiple Access with Side Information for V2X Communications

Zhenhui Situ[1] and Ivan Wang-Hei Ho[1,2(✉)]

[1] Department of Electronic and Information Engineering,
The Hong Kong Polytechnic University, Hung Hom, Hong Kong
z.situ@connect.polyu.hk, ivanwh.ho@polyu.edu.hk
[2] The Hong Kong Polytechnic University Shenzhen Research Institute,
Shenzhen, China

Abstract. We study a vehicle-to-everything (V2X) communication scenario where multiple vehicles coming from different road segments converge at a road junction and exchange their information via a road side unit. The high-mobility of vehicles determines that the communication is time-critical. If conventional orthogonal multiple access (OMA) is applied, not only the orthogonal resource allocation but also the scheduling overheads will incur significant delay. In addition, orthogonal domain may not be identified within short contact time among vehicles and the road-side unit. In contrast, non-orthogonal multiple access (NOMA) can provide low-delay and reliable communication by exploiting the overlapped or collided signals. In this paper, we investigate the application of NOMA with side information in V2X communications as the Non-orthogonal V2X (NO-V2X) scheme. NO-V2X takes the advantage of side information and physical-layer network coding (PNC) to increase the decoding success rate in the uplink phase and to reduce the required transmission power in the downlink phase. Our simulation results show that NO-V2X outperforms OMA and the conventional NOMA with successive interference cancellation (SIC).

Keywords: Non-orthogonal multiple access · V2X communication
Side information · Physical-layer network coding
Network-coded multiple access

1 Introduction

The third generation partnership project (3GPP) group standardized the initial Cellular Vehicular-to-Everything (C-V2X) communication in Release 14 [1]. The document justifies that the world-wild research on connected vehicles (e.g. CCSA in China) shows the market requirement on Long Term Evolution (LTE) based V2X communication [2]. To respond to such situation, the standard proposes LTE-based V2X communication and offers three types of communications:

ⓒ ICST Institute for Computer Sciences, Social Informatics and Telecommunications Engineering 2018
P. H. J. Chong et al. (Eds.): SmartGIFT 2018, LNICST 245, pp. 133–144, 2018.
https://doi.org/10.1007/978-3-319-94965-9_14

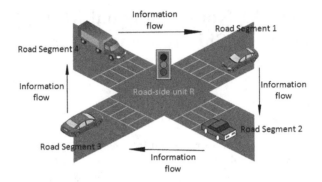

Fig. 1. V2X communication at a crossroad.

vehicle-to-vehicle (V2V), vehicle-to-infrastructure/network (V2I/N) and vehicle-to-pedestrian (V2P) communications. Soon after, the 5G Automotive Association (5GAA) employed C-V2X for safety and cooperative driving in its white paper [3]. V2X is expected to provide real-time/low-latency, high-reliable communications in a dense moving environment due to the characteristic of vehicles.

Figure 1 illustrates a common V2X communication scenario. Vehicles from different directions converge at a road junction, and they have different destination road segments. In such a short duration, each vehicle needs to transmit its self-information (e.g. traffic information of the road segment that it has traveled) to another vehicle (e.g. traffic information of the destination road segment). We consider the following transmission pattern: $1 \rightarrow 2 \rightarrow 3 \rightarrow 4 \rightarrow 1$, as illustrated in Fig. 1. The arrows represent the direction of information flow. Since the radio channels between any two vehicles are obstructed by buildings or other obstacles, all vehicles need the road-side unit, R, to relay the self-information.

Based on orthogonal multiple access (OMA), the current LTE allocates spectrum resources to users in an orthogonal manner. For the case in Fig. 1, vehicles transmit data in either different time or frequency to the relay in the uplink phase, then the relay forwards the data in the same manner in the downlink phase. It is uncertain if OMA can meet the requirement of V2X communication. Considering a dense vehicular network in urban area, OMA needs to perform orthogonal resource allocation and requires overheads to schedule multiple users, which leads to increasing delay as the number of users increases. In addition, orthogonal domain may not be identified within short contact time among vehicles and the road-side unit. For the case in Fig. 1, OMA with 4-user access generates at least four times latency compared with single-user access. By exploiting the overlapped or collided signals, non-orthogonal multiple access (NOMA) [4] is considered as a promising candidate to achieve low-latency and high-reliability in V2X communication.

The concept of NOMA has been proposed and studied for a long time [5,6], but the research on the practical application is relatively new and shoot up in recent years. NOMA allows multiple users to be served by one transceiver simultaneously at either the uplink and downlink phase. By sharing the spectrum resources at the same time, NOMA can boost the throughput and reduce the latency significantly, especially when the number of access users is large [7]. For the same case in Fig. 1, vehicles from different directions transmit data at the same time and frequency to the relay, the relay decodes the individual information and forwards it to all end nodes simultaneously. Thus, the delay of NOMA with 4-user access is the same as that of single-user access theoretically. To cope with the co-channel inference caused by spectrum sharing, various multi-user detection (MUD) techniques such as successive interference cancellation (SIC) [8] have been proposed. In this paper, we investigate the feasibility of applying NOMA in V2X communications. If the application is feasible, we would like to see if the performance can be further improved with respect to the characteristic of V2X communications.

Generally, a SIC-based NOMA receiver decides the decoding order according to the received power level. The user with the strongest power is decoded first and the signals from other users are regarded as noise. After one user is decoded, the corresponding signal is removed and the next strongest user is decoded. It turns out that the power difference among multi users is essential for SIC receivers. For 2-users access, when the power levels of two users are close, the decoding signal-to-noise ratio (SNR) for the strongest user is lower than 0 dB. Under such situation, the SIC receiver may fail to decode the strongest user due to the low SNR, and the other user cannot be decoded as well since the interference from the strongest user cannot be removed. Power control is simple in the downlink phase but difficult in the uplink phase, a common way to guarantee the power difference in the uplink phase is user grouping. This method requires the channel state information (CSI) to pair the strong user and the weak user [9].

Recently, several works on NOMA in V2X communications have been published [10–12], the three schemes are based on SIC. Specifically, [10,11] focus on scheduling design and resource allocation algorithm to maximize the decoding success rate and throughput, and [12] investigated the use of side information, which is used for interference cancellation for the strong user, to enhance the transmission of the weak user. Similar idea that exploits side information has been studied in previous works [13], the difference is that [12] studied the performance under multiple-input multiple-output spatial modulation (MIMO-SM) and V2V networks.

Different from static networks, the high-mobility of vehicles makes power control and user grouping in the uplink phase challenging, and this may destroy the time-critical transmission in Fig. 1. In this work, we investigate the possibility of removing power control and user grouping in the uplink phase. Specifically, all end nodes with random CSIs are allowed to transmit data at the same power level simultaneously, thus the overhead and delay can be reduced. In the downlink phase, we exploit the side information used for SIC to further improve the data exchange shown in Fig. 1. Different from the previous use of side information which simply re-transmit the side information to the weak user, we exploits the side information according to the pattern of V2X communications. Overall, this paper proposes a novel NO-V2X architecture that exploits the side information (e.g. self-information in the uplink phase and information used for SIC in the downlink phase) to achieve low-latency and reliable communication, the salient features of NO-V2X are as follows:

1. In the uplink phase, NO-V2X applies Network-Coded Multiple Access (NCMA) [14] and side information to increase the decoding success rate. The proposed NO-V2X outperforms SIC-based NOMA and OMA under independent Rayleigh fading channels.
2. In the downlink phase, benefit from the side information, NO-V2X applies Physical-layer Network Coding (PNC) to encode the transmitted data and the amount of data can be reduced by $\frac{1}{\# \text{ users}}$. In other word, the required transmission power can be reduced for achieving the same throughput and delay as conventional NOMA.

The remainder of this paper is organized as follows. Section 2 introduces the system models. After that, the proposed NO-V2X is discussed in Sect. 3. NO-V2X can be divided into the uplink and downlink phases, and they are described in detail and evaluated with simulation results. Finally, Sect. 4 concludes the paper.

2   System Models

Before introducing the proposed architecture, this section provides the details with respect to the system models. In this paper, we consider the V2X communication scenario as shown in Fig. 1.

2.1   NOMA Uplink

In the uplink phase, K vehicles from different directions upload the self-information to the relay R. The received signal at R is

$$y(t) = \sum_{k=1}^{K} h_k(t)x_k(t) + n(t) \qquad (1)$$

where t is the time index, $h_k(t)$ is the channel gain, $x_k(t)$ denotes the transmitted data, and $n(t)$ is the complex white Gaussian noise (WGN) with zero mean and variances N_o. The transmitted data undergo independent radio channel and the corresponding channel gain can be written as the product of slow fading and fast fading:

$$h_k(t) = h_k^{slow}(t)h_k^{fast}(t) \qquad (2)$$

The slow fading is simplified as one and the fast fading is modeled as independent Rayleigh fading with expectation:

$$E(|h_k^{fast}(t)|^2) = 1 \qquad (3)$$

2.2   NOMA Downlink

In the downlink phase, SIC-based NOMA assigns different power levels to users according to the CSI. The transmitted signal from R is expressed as:

$$x(t) = \sum_{k=1}^{K} \alpha_k x_k(t) \tag{4}$$

The users with the worst channel (e.g. user $k = \arg\min_k |h_k(t)|^2$) is assigned with the maximum power, other users need to decode the data of user $k = \arg\min_k |h_k(t)|^2$ and remove it with SIC before decoding the target data. To guarantee the fairness among all users, we assign the power coefficients α_k to ensure that all users share the same capacity:

$$C_k = log(1 + \frac{|\alpha_k h_k|^2}{N_o + (\sum_{\{j||h_j|>|h_k|\}} |\alpha_j|^2)|h_k|^2})$$

$$C_k = C_j, \forall k \neq j \tag{5}$$

$$\sum_{k=1}^{K} \alpha_k^2 = 1$$

2.3   Physical-Layer Network Coding (PNC) and Network-Coded Multiple Access (NCMA)

PNC is widely studied to boost the performance of two-way relay channel (TWRC) [15]. Consider a similar network in Fig. 1 with two users only. In this case, PNC decodes the XOR-ed output of two users $x_R = x_1 \oplus x_2$ instead of the individual information and broadcasts it in the downlink phase. One significant feature is that PNC works well under the power-balanced situation. Thus, PNC is regarded as NOMA and a new MUD architecture named Network-Coded Multiple Access (NCMA) was proposed in [14]. Consider a 2-user access case with $|h_1| > |h_2|$, NCMA decodes not only the individual information but also the XOR-ed output. SIC-based NOMA detects the strongest user while regarding the signals from other users as noise. The SIC decoder calculates the likelihood function of the data $x_1(t)$, namely the probability of $x_1(t)$ given the received signal $y(t)$, as

$$p(x_1(t)|y(t)) \propto \exp\{-|y(t) - h_1(t)x_1(t)|^2/(N_o + |h_2(t)|^2)\} \tag{6}$$

where the expected power of transmitted data $E(|x_2|^2)$ is normalized as one. In PNC, the detection is different. Specifically, the PNC decoder first calculates the likelihood function of the data pair $(x_1(t), x_2(t))$ as

$$p(x_1(t), x_2(t)|y(t)) \propto \exp\{-|y(t) - h_1(t)x_1(t) - h_2(t)x_2(t)|^2/N_o\} \tag{7}$$

And the likelihood function of individual information is calculated as:

$$p(x_1(t)|y(t)) = \sum_{x_2(t)} p(x_1(t), x_2(t)) \tag{8}$$

Similarly, the likelihood function of XOR-ed data is equivalent to:

$$p(x_1(t) \oplus x_2(t)|y(t)) = \sum_{x_1(t) \oplus x_2(t)} p(x_1(t), x_2(t)) \qquad (9)$$

After the individual and XOR-ed data detection, the receiver may decode partial data as illustrated in Table 1. The black ticks shown in the table denote the successfully received packets (e.g., with CRC checking). As can be seen, three out of six packets of x_1 are received while only two packets of x_2 are correct. In the meanwhile, $x_1 \oplus x_2$ is also decoded and two packets pass the checking. In NCMA, bridging on PHY and MAC layers is performed:

PHY-layer Bridging: In time slot three and five, both x_1 and $x_1 \oplus x_2$ are decoded, in this case, x_2 can also be decoded with $x_2 = x_1 \oplus x_1 \oplus x_2$. Thus, $x_1 \oplus x_2$ can be used to recover x_1 or x_2 via PHY-layer bridging, and the data recovered by PHY-layer bridging are denoted with blue circles.

MAC-layer Bridging: Applying an erasure channel code (e.g., the Reed-Solomon (RS) code), one frame is partitioned into multiple packets. We assume the use of a (6, 3) RS code, and thus one frame can be recovered once three packets are received. In Table 1, the packets of x_1 at time slots one, four and six can be recovered with RS code, and we denotes them with red triangles. Then, we can decoded the lone $x_1 \oplus x_2$ packets at time slot one and four accordingly.

Through performing the PHY-layer and MAC-layer bridging, the PNC decoder can be used to enhance the MUD decoder, and NCMA has been theoretically and practically verified to be a feasible receiver for power-balanced NOMA [16]. However, the goal of NCMA is to decode the individual information of all users under the access point (AP) mode, which may be an overkill for the V2X scenario considered as the vehicles are supposed to have side information.

Table 1. NCMA decoding

Time slot	x_1	x_2	$x_1 \oplus x_2$
1	△	√	
2	√		
3	√	○	√
4	△	√	
5	√	○	√
6	△		

3   NO-V2X

In the introduction section, a question regarding the feasibility of NOMA in V2X communications is raised. We give the positive response to the question by exploiting the usage of side information in this section. Our study is divided into the uplink and the downlink phases, and we will show how NOMA based V2X communication benefits from side information in the two phases. To begin with, we study the feasibility of allowing all users with random CSIs to transmit data in the same power level.

3.1 NO-V2X Uplink

As discussed above, the key bottleneck to apply NOMA in V2X communications is the delay caused by complicated scheduling and resource allocation. Unlike the previous NOMA architectures that employ either power control or user grouping, the proposed NO-V2X allows all users with diverse CSIs to transmit data in the same power level simultaneously in the uplink phase, thus the requirement of low-latency can be achieved. PNC is verified to provide good performance under the power-balanced situation [14,16], thus is exploited in NCMA to boost the MUD throughput.

In this work, we further investigate the direct usage of the XOR-ed output of two packets (e.g., $x_1 \oplus x_2$). For the V2X communication shown in Fig. 1, user 1 requires the information from direction 4 (i.e., x_4). Inspired from PNC, relay R can either transmit x_4 or $x_1 \oplus x_4$ to user 1 in the downlink phase since the user can recover $x_4 = x_1 \oplus x_4 \oplus x_1$ with the side information x_1. Similarly, the decoded XOR-ed output of other packets can be used in the downlink phase since the end node can recover the target data with side information. Previous work [16] has proved the high success rate of PNC decoders under power-balanced situation. We further evaluate the proposed NO-V2X under random power levels. Specifically, independent Rayleigh fading is utilized to model the channels for all users.

Figures 2 and 3 show the throughput and delay of the V2X network illustrated in Fig. 1. Three types of decoders are simulated: OMA, SIC-based NOMA and NO-V2X. As can be seen, the two NOMA methods outperform the OMA method, the gap reaches the highest in the high SNR regime. Notice that the maximum throughput is 4 packets per time slot and the minimum delay is 0.25 time slot per packet when there is no error. But the SIC decoder can only achieve the throughput of less than 2.5 packets per time slot. The reason is that the success of SIC decoding strongly depends on the power difference among users. When there is no power control, SIC decoder cannot guarantee reliable communication, thus is not suitable for such V2X communication scenario in Fig. 1. By contrast, the NO-V2X decoder shows high throughput and low delay under random power levels. Furthermore, NO-V2X reaches the maximum throughput and minimum delay at 20 dB. Actually, in the low SNR regime (e.g., 4 dB), NO-V2X already provides more than 60% throughput improvement compared to the other two methods. Therefore, NO-V2X appears to be the solution to satisfy the requirement of random power levels in NOMA, and enables robust time-critical V2X communications.

In addition, the 2-user and 3-user uplink cases are shown in Figs. 4 and 5 to compare the performance of the three methods under different number of

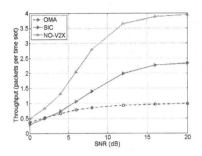

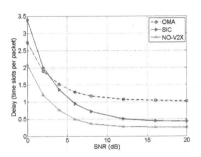

Fig. 2. Throughput of 4-user uplink.   **Fig. 3.** Delay of 4-user uplink.

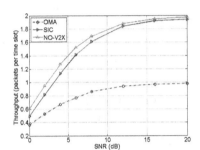

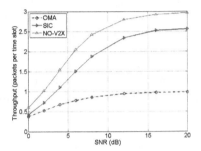

Fig. 4. Throughput of 2-user uplink.   **Fig. 5.** Throughput of 3-user uplink.

access users. The 2-user case is equivalent to the TWRC network and the 3-user network is performing data exchange at a T junction. The two simulations draw the same conclusions: (1) NO-V2X provides the best performance, followed by SIC, and the two NOMA methods outperforms the OMA method; and (2) NO-V2X can reach the maximum throughput at around 20 dB SNR. The key difference among Figs. 2, 4 and 5 is that the gaps between any two algorithms become larger as the number of access users increases. Therefore, NO-V2X is especially suitable for practical dense V2X communications where the number of connected vehicles is large.

3.2   NO-V2X Downlink

The previous subsection studies the feasibility of NOMA in V2X communications. The study shows that NO-V2X offers the positive response via using NCMA and side information in the uplink phase. In this subsection, we investigate the downlink phase. For simplification, data from all users are assumed to be successfully decoded in the uplink phase. Different from the uplink phase, power control is simple in the downlink phase and is widely studied and utilized in NOMA systems. Inspired from the use of side information in coordinated direct and relay transmission, this work exploits the information for interference cancellation in the downlink phase. Instead of broadcasting the individual data,

NO-V2X transmits the XOR-ed data in the downlink phase. The algorithm of the NO-V2X downlink is illustrated in Algorithm 1.

Algorithm 1. Pseudocode for NO-V2X downlink

Encode data transmitted from user i to j as $x_i \oplus x_j$

Eliminate the encoded data sent to the strongest user (i.e., the user with the best channel)

According to (5), assign the power coefficient α_k to other users

Broadcast the superimposed signal according to (4)

for All users j **do**

    Decode the XOR-ed data $x_i \oplus x_j$ with NOMA receiver, for the strongest user k, the target XOR-ed data is equivalent to the XOR-ed output of all XOR-ed data

    Recover the target data $x_i = x_i \oplus x_j \oplus x_j$ with side information x_j

end for

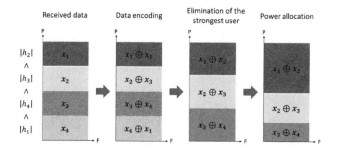

Fig. 6. An example of NO-V2X downlink.

For the data x_i transmitted from user i to j, conventional NOMA systems broadcast it directly in the downlink phase. But it is possible to transmit $x_i \oplus x_j$ instead of x_i as user j can recover x_i with the side information x_j. Let us assume the CSIs in Fig. 1 as $|h_1| > |h_4| > |h_3| > |h_2|$, the relay encodes the data as shown in Fig. 6. The first step is to encode the data x_i transmitted from user i to j as $x_i \oplus x_j$, e.g., the data x_1 sent to user 2 is encoded as $x_1 \oplus x_2$. The next step is to remove the data sent to the strongest user since the strongest user can recover the target data from the side information sent to other users. For instance, user 1 undergoes the best channel in Fig. 6, thus the encoded data sent to user 1 $x_4 \oplus x_1$ are deleted. But user 1 can recover the target XOR-ed data as $x_4 \oplus x_1 = (x_1 \oplus x_2) \oplus (x_2 \oplus x_3) \oplus (x_3 \oplus x_4)$. The final step is to assign the power coefficients α_k to other users according to (5), the user with better channel is allocated with less power. It can be seen in Fig. 6 that user 2 with the worst channel is assigned with the largest power. After receiving the encoded data, each user recovers the target data with the side information. By exploiting the side information, NO-V2X can reduce the amount of transmitted data by one user in the downlink phase.

Figures 7 and 8 show the throughput and delay for the V2X network illustrated in Fig. 1. In the low SNR regime, the OMA method shows better performance while the two NOMA methods provides low throughput and high delay. The reason is that the limited transmission power provided by one relay can hardly satisfy the throughput requirement of multiple users. In fact, the two NOMA algorithms can provide better overall throughput by applying other power allocation algorithms such as water-filling algorithm [17]. The goal of this work is to guarantee fairness, thus the power allocation algorithm in (5) assigns the same capacity to all access users. When the SNR is around 10 dB, the two NOMA methods reach similar throughput and delay as the OMA method. The advantage of NO-V2X comes from the side information, and thus the data of the strongest user can be removed. In the 4-user access case, NO-V2X reduces $\frac{1}{4} = 25\%$ data in the downlink phase. This results in around 0.5 throughput improvement compared with SIC-based NOMA. When the SNR is higher than 10 dB, NO-V2X provides more than 2 dB SNR gain compared with the other two methods.

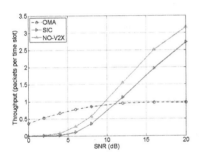

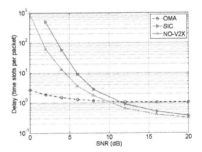

Fig. 7. Throughput of 4-user downlink.   **Fig. 8.** Delay of 4-user downlink.

Similar to the uplink phase, the 2-user and 3-user cases are also considered and simulated, the results are shown in Figs. 9 and 10. From the two simulations, we can observe that the OMA provides better performance compared with SIC in the low SNR regime. By exploiting the side information, NO-V2X reaches the same throughput as the SIC method but reduces the amount of transmitted data. The percentages of reduced data in the two cases are $\frac{1}{2} = 50\%$ and $\frac{1}{3} \approx 33\%$, respectively. In the 2-user access case, the relay transmits the XOR-ed output of the data received from the two users and no data overlap in the power domain. This case is equivalent to PNC in TWRC and the two users can recover the target data via the same XOR-ed data. Under such situation, NO-V2X outperforms OMA, even though in the low SNR regime. One interesting point is that NO-V2X shows larger throughput gap compared with SIC when the number of access

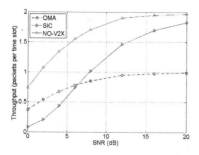

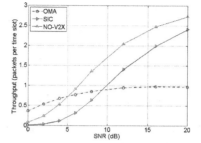

Fig. 9. Throughput of 2-user downlink.

Fig. 10. Throughput of 3-user downlink.

user increases in the uplink phase, but this trend is inverse in the downlink phase. However, the utilization of other power allocation algorithms is possible to improve the performance in the downlink phase. NO-V2X improves the downlink phase by reducing the amount of transmitted data, and it is compatible to any power control algorithms.

4   Conclusion and Future Work

In this paper, we have investigated the feasibility of NOMA in V2X communications. By studying a time-critical V2X communication, we proposed a NO-V2X architecture that exploits the use of side information to boost the throughput and to reduce the delay in both the uplink and downlink phases. The paper has studied three problems: (1) the feasibility of allowing all users with random CSIs to transmit data in the same power level in the uplink phase is verified with the support of PNC. We show that the integration of MUD and PNC makes time-critical V2X communications possible; (2) Different from the conventional individual user detection, NO-V2X exploits the network-coded side information and further improve the decoding success rate in the uplink phase; and (3) the required transmission power in the downlink phase can be reduced since NO-V2X removes the data sent to the strongest user. Overall, our simulation results verify that NO-V2X outperforms OMA and SIC-based NOMA. Specifically, for the 4-user access case, NO-V2X provides more than 60% throughput enhancement when the SNR is higher than 4 dB in the uplink phase and more than 2 dB SNR gain when the SNR is higher than 10 dB in the downlink phase. However, there are still many extensions that can be done. For instance, besides the circular transmission pattern considered, other cases with a random data demand will be studied. In addition, we studied the uplink and downlink phases separately in this paper, joint consideration of both phases will be our next step.

Acknowledgment. The work of I. W.-H. Ho is partially supported by the Early Career Scheme (Project No. 25200714) established under the University Grant Committee (UGC) of the Hong Kong Special Administrative Region (HKSAR), China; and by The Hong Kong Polytechnic University (Projects G-YBK6, G-YBR2, G-YBXJ).

The work of Z. Situ is partially supported by the National Natural Science Foundation of China (Project No. 61401384).

References

1. 3rd Generation Partnership Project: Release 14 (2017). http://www.3gpp.org/release-14
2. Araniti, G., Campolo, C., Condoluci, M., Iera, A., Molinaro, A.: LTE for vehicular networking: a survey. IEEE Commun. Mag. **51**(5), 148–157 (2013)
3. 5G Automotive Association, et al.: The case for cellular V2X for safety and cooperative driving. 5GAA Whitepaper, November 2016
4. Saito, Y., Kishiyama, Y., Benjebbour, A., Nakamura, T., Li, A., Higuchi, K.: Non-orthogonal multiple access (NOMA) for cellular future radio access. In: IEEE Vehicular Technology Conference (VTC Spring), pp. 1–5 (2013)
5. Verdu, S.: Multiuser Detection. Cambridge University Press, Cambridge (1998)
6. Wang, X., Poor, H.V.: Wireless Communication Systems: Advanced Techniques for Signal Reception. Prentice Hall Professional, Upper Saddle River (2004)
7. Xu, C., Hu, Y., Liang, C., Ma, J., Ping, L.: Massive MIMO, non-orthogonal multiple access and interleave division multiple access. IEEE Access **5**, 14 728–14 748 (2017)
8. Tse, D., Viswanath, P.: Fundamentals of Wireless Communication. Cambridge University Press, Cambridge (2005)
9. Ding, Z., Fan, P., Poor, H.V.: Impact of user pairing on 5G nonorthogonal multiple-access downlink transmissions. IEEE Trans. Veh. Technol. **65**(8), 6010–6023 (2016)
10. Di, B., Song, L., Li, Y., Li, G.Y.: Non-orthogonal multiple access for high-reliable and low-latency V2X communications in 5G systems. IEEE J. Sel. Areas Commun. **35**(10), 2383–2397 (2017)
11. Qian, L.P., Wu, Y., Zhou, H., Shen, X.: Dynamic cell association for non-orthogonal multiple-access V2S networks. IEEE J. Sel. Areas Commun. **35**(10), 2342–2356 (2017)
12. Chen, Y., Wang, L., Ai, Y., Jiao, B., Hanzo, L.: Performance analysis of NOMA-SM in vehicle-to-vehicle massive MIMO channels. IEEE J. Sel. Areas Commun. **35**(12), 2653–2666 (2017)
13. Kim, J.-B., Lee, I.-H.: Non-orthogonal multiple access in coordinated direct and relay transmission. IEEE Commun. Lett. **19**(11), 2037–2040 (2015)
14. Lu, L., You, L., Liew, S.C.: Network-coded multiple access. IEEE Trans. Mob. Comput. **13**(12), 2853–2869 (2014)
15. Liew, S.C., Zhang, S., Lu, L.: Physical-layer network coding: tutorial, survey, and beyond. Phys. Commun. **6**, 4–42 (2013)
16. Pan, H., Lu, L., Liew, S.C.: Practical power-balanced non-orthogonal multiple access. IEEE J. Sel. Areas Commun. **35**(10), 2312–2327 (2017)
17. Kobayashi, M., Caire, G.: An iterative water-filling algorithm for maximum weighted sum-rate of Gaussian MIMO-BC. IEEE J. Sel. Areas Commun. **24**(8), 1640–1646 (2006)

Connecting Makaraka - A Case Study to Provide Connectivity in the Rural Area of New Zealand

Syeda Kanwal Zaidi[✉], Ali Abdul Adheem, Syed Faraz Hasan, and Xiang Gui

Massey University, Palmerston North, New Zealand
{k.zaidi, a.aldheem, f.hasan, x.gui}@massey.ac.nz

Abstract. Broadband availability is an important asset for deriving change in the societal and economic development of a country. New Zealand's rural areas have limited broadband connectivity and there are some beautiful tourist destinations which do not have broadband availability on site. In this paper, we aim to provide a broadband connectivity model for basic and high speed data services for the area of Makaraka in the Gisborne region of New Zealand. The model proposes the placement of the access points, their channel selection scheme, transmit power requirements and appropriate installation height in detail for different frequency bands and data rate requirements. It is evident from the simulated results that outdoor area of Makaraka is able to get broadband connectivity which will help NZ to drive towards the theme of highly-connected society.

Keywords: Broadband communication · Internet · Data connectivity

1 Introduction

The ever-developing digital world is significantly affecting many aspects of New Zealanders and their lives. Information Communication Technology has become the vital part of the businesses, education sector and health care. In line with the ICT, there is an increasing usage of the broadband at every stage for example from basic data services (web and email) to high quality video streaming and gaming. Therefore, Government of New Zealand is aiming to provide broadband and cellular coverage to majority of the areas for making New Zealand a leading digital and highly connected nation. Rural areas of New Zealand have the low population density and network operators have not invested initially in providing broadband coverage to these areas. According to a study in [1], many of the rural areas of Australia and nearby countries also face the broadband un-availability and dis-connectivity issues.

Recently Ministry of Business, Innovation and Employment has initiated Rural Broadband Initiative to deliver better and improved wireless broadband services to the rural and urban areas of New Zealand. Several initiatives have been taken by the higher authorities for example Digital Economy, Mobile Black Spot fund, Ultra-fast Broadband Initiative and Rural Broadband Initiative to progress in the information and communication technology [2]. Availability of broadband connection may be useful in

P. H. J. Chong et al. (Eds.): SmartGIFT 2018, LNICST 245, pp. 145–154, 2018.
https://doi.org/10.1007/978-3-319-94965-9_15

the foundation of many more applications other than Internet and data-browsing, for example Indoor Localization, health management and improved surveillance systems [3, 4]. The region of Gisborne comes in the plan to receive new Ultra-Fast Broadband, Rural Broadband and Mobile Blackspot programme coverage. However, there are some specific areas within the Gisborne region which need further detail and planning. Makaraka is a small area within Gisborne with very low population and doesn't have broadband connectivity across it. In this paper, we aim to propose an initial broadband design for Makaraka area in the Gisborne region to provide basic and high speed connectivity. We propose the coverage plan and design considerations to deploy the broadband network successfully within the region. We also provide the coverage map and limitations for two connectivity schemes that is basic and high speed connectivity under two different frequency considerations, 2.4 GHz and 5 GHz.

1.1 Background

The existing outdoor area of Makaraka has no broadband coverage and there is only one fiber optic cable line present at Makaraka School. There is a need to provide Wireless Internet coverage based on advanced technologies which should be aligned to standard Wi-Fi systems. There is a need for reliable and constant Internet connection while moving around the Makaraka central area. The broadband access system is expected to cover the outdoor area of Makaraka stretched over 3.7 × 3 km square area (approx.). However, because of the low population density at the borders of the area, this paper focuses on providing the coverage in the central densely populated area of Makaraka.

1.2 Objectives

The objectives of this paper is to provide a proposal for installing an outdoor Wi-Fi system for Makaraka area. The scope of the paper includes the planning and designing of the outdoor Wi-Fi Internet Access System for Makaraka. The equipment supply, delivery to site, installation, testing and commissioning of the outdoor Wi-Fi system for Makaraka area are out of scope of this paper. The ultimate goals for the project are to:

- Identify the best fit equipment for installing the outdoor Wi-Fi system for Makaraka area.
- Adopt the latest wireless standard of dual-band implementations (2.4 GHz and 5 GHz) with multiple-input and multiple-output (MIMO) technology.
- Provide better wireless performance, reliability and availability (a better user experience).

2 Network Design

We propose a wireless mesh network based design for providing broadband connectivity in the specified area of Makaraka. The two important design components of the network are Mesh Gateway and Mesh Nodes, explained below.

Mesh Nodes: Mesh Nodes are the access points responsible to provide Wi-Fi coverage to the users in the area and establish a wireless backhaul link to the backbone network. The primary focus of these nodes is to provide 802.11 access capabilities and forward the data of the users to and from the Internet. They also enforce Quality of Service rules for the data traffic and are the end points for the subscribers to get internet access.

Mesh Gateway: Mesh Gateway or root access point is responsible to forward the traffic between Mesh nodes and the central backhaul network at Makaraka School. A mesh gateway is responsible to handle the data from different mesh nodes. The nodes automatically select the shortest path to gateway based on the dedicated algorithm.

Figure 1 below shows the wireless mesh network model for the Makaraka Area, which involves a Mesh Gateway (Wireless Controller) and several Mesh Nodes (Wireless Mesh Access Points).

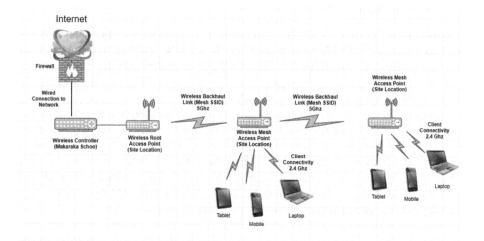

Fig. 1. Proposed network model

3 Design Considerations

Following are some important design considerations for the Wireless Network of Makaraka.

- Maximum number of client devices on each Access Point (AP): The number of clients connected to each access points may vary depending upon the type of application used by the clients and the type of the access point. For Basic Connectivity (includes Web, Email, Multimedia), it is recommended to have 40 to 60 users for a 2×2 MIMO AP and 60–80 users for a 3×3 MIMO AP.
- Transmission Power Level of each AP: 18 dBm.

- Channel Reuse Pattern: Fixed or Flexible (Channels should not overlap with neighboring AP).
- Received Signal Strength Indicator (RSSI) and Signal to Noise Ratio (SNR) for data Services: Minimum RSSI: -70 dBm, SNR: 20 dB or higher.
- Data + Multimedia: Minimum RSSI: -67 dBm, SNR of 23 dB or higher.

3.1 Wireless Mesh Access Points

We consider the network to be Wireless Mesh Network where APs communicate among themselves and back to root access point wirelessly over a radio backhaul. All APs in the network mesh have fixed configuration for the wireless mesh backhaul communication. We consider that APs should use a protocol to decide the best path through the other mesh APs to the main AP and also, AP should be dual-band (2.4 GHz and 5 GHz) with MIMO technology.

3.2 Wireless Root Controller

In our proposed design, all of the APs need to be connected to the central Wi-Fi controller wirelessly. The Wireless Root Controller is to be places in Makaraka School at a suitable location. We propose to connect Wi-Fi controller with the wired network for Internet connectivity and manage it centrally. The root AP connected to the controller is able to support all other mesh AP.

We consider to locate the Root Access Point on a tall building or tower at Makaraka School. However, the Mesh Access Point locations are short building tops or street poles. In our design, it is recommended to place AP at a distance between 200 m to 400 m if the backhaul mesh link is at 5 GHz. An outdoor AP may serve a client at a distance of 300 to 500 m. There should be maximum 3 to 4 hops between root AP and mesh AP. More hops can be supported but this is the recommended figure. Client are typically laptops, personal cell phones, or hand held devices.

4 Proposed Network Design

In this section, we explain the complete network design to provide basic or high speed connectivity in the Makaraka area based on dual frequency standards. The model has been proposed based on the simulation results of Aerohive online network development tool [5]. There can be modifications made to this design to provide better connectivity and reliability based on the number of users in the area. Table 1 highlights the access point locations for basic and high speed connectivity under 2.4 GHz frequency band. The expected data rate that can be achieved in the specified region which lies in the range of 78 Mbps to 174 Mbps for basic connectivity and 104 Mbps or more for high speed connectivity.

4.1 Basic Connectivity- 2.4 GHz

Figures 2 and 3 below show the received signal strength indicator (RSSI) Heat map when five access points are placed at the positions specified in Table 1. The channel selection is done manually by assigning non-over lapping channels to avoid interference.

Table 1. Access point locations (2.4 GHz)

Type of connectivity	Location of the access point	Channels assigned	Distance to neighboring AP (meters)
(Basic connectivity)	Makaraka school	11	440
	Rose land tavern	1	394
	AGRIplus	6	357
	Intersection of main road and Paker Ln	11	471
	Motor camp entrance (Gate 3)	1	
High speed connectivity	Makaraka school	6	352
	117 main road	11	326
	Makaraka veges	1	336
	67 main road	6	189
	Intersection of main road and Paker Ln	11	331
	35 main road	6	

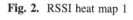

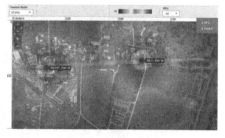

Fig. 2. RSSI heat map 1

Fig. 3. RSSI heat map 2

4.2   High Speed Connectivity- 2.4 GHz

Figures 4 and 5 below show the RSSI Heat map when six access points are placed at the positions specified in Table 1 for high speed connectivity at 2.4 GHz band.

Table 2 highlights the access point locations for basic and high speed connectivity under 5 GHz frequency band.

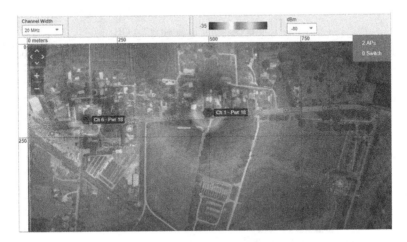

Fig. 4. High- speed connectivity (2.4 GHz) - RSSI heat map 1

Fig. 5. High- speed connectivity (2.4 GHz) - RSSI heat map 2

Table 2. Access point locations (5 GHz)

Type of connectivity	Location of the access point	Channels assigned	Distance to neighboring AP (meters)
Basic connectivity	Makaraka school	112	338
	117 main road	108	376
	Makaraka veges	100	285
	67 main road	104	189
	Main/Paker Ints.	165	331
	35 main road	100	
High speed connectivity	Makaraka School	112	338
	117 main road	108	376
	Makaraka veges	100	285
	67 main road	104	189
	Main/Paker Ints.	165	331
	35 main road	100	

4.3   Basic Connectivity- 5 GHz

Figures 6 and 7 below show the RSSI Heat map when five access points are placed at the positions specified in Table 2.

Fig. 6. Basic connectivity (5 GHz) - RSSI heat map 1

Fig. 7. Basic connectivity (5 GHz) - RSSI heat map 2

4.4   High Speed Connectivity- 5 GHz

Figures 8 and 9 below show the RSSI Heat map when six access points are placed at the positions specified in Table 2.

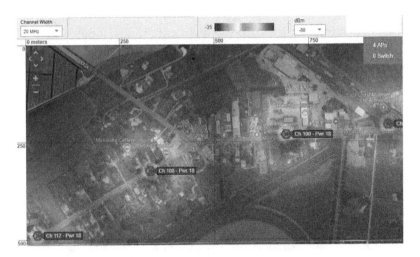

Fig. 8. High-speed connectivity (5 GHz) - RSSI heat map 1

Fig. 9. High-speed connectivity (5 GHz) - RSSI heat map 2

5   Wireless Network Requirements

In order to connect to Wireless Network, users are required to open SSID with a web portal so that a secure connection can be established. The client types for wireless network are 802.11a/b/g/n smart devices from locals and visitors. Client applications and network use is limited to Internet access for web browsing and email. The network is expected to support mobility that is users might move around while using their devices and will get coverage from multiple APs on the go. With the help of the proposed design, we can say that

- The Root Access Point to Mesh Access Point ratio is 10 Mesh Access Points per Root Access Point.
- The maximum recommended distance between two APs is 500 meters.
- An area of around one square km (1 km^2) comprises of three cells minimum and can be covered with two or more hops.

6   Limitations of the Model

- The number of users within the specified coverage area of an AP are not known. It is expected that with the correct number of users, the requirement of the access points might increase which may increase the cost of the network.
- The attenuation and noise present at the specified location of APs is not known. With practical Radio frequency spectrum analyser and calculation of attenuation and noise at the locations of Access points, the coverage and rate parameters may vary.
- The total cost of the model may vary if more number of access points need to be deployed to meet the data demand for more number of users.

7   Conclusion

In this paper, we have proposed an outdoor broadband coverage model for Makaraka area in the region of Gisborne. The proposed model provides basic and high speed connectivity to the central area under two different frequency schemes. We have highlighted the physical installation locations along with the expected throughput that can be achieved. A future extension of this work is the testing in the real environment of Makaraka which can help in further improvement of the design.

Acknowledgement. This work has been funded by Massey University Research Fund, 2017/2018. The authors acknowledge the generous support of Manu Caddie, Maurice Alford and Hari Shankar for their valuable feedback on this work.

References

1. Carrillo, D., Seki, J.: Rural area deployment of internet of things connectivity: LTE and LoRaWAN case study. In: IEEE XXIV International Conference on Electronics, Electrical Engineering and Computing (INTERCON), pp. 1–4. IEEE (2017)
2. Ministry of Business Innovation and Employment. http://www.mbie.govt.nz/info-services/infrastructure-growth/national-broadband-initiatives
3. Martelli, T., Pastina, D., Colone, F., Lombardo, P.: Enhanced WiFi-based passive ISAR for indoor and outdoor surveillance. In: Radar Conference (RadarCon), pp. 0974–0979. IEEE (2015)
4. Wang, J., Tan, N., Luo, J., Pan, S.J.: WOLoc: WiFi-only outdoor localization using crowdsensed hotspot labels. In: Proceedings of the IEEE INFOCOM (2017)
5. Aerohive Networks. https://cloud.aerohive.com

A New Energy Efficient Big Data Dissemination Approach Using the Opportunistic D2D Communications

Ambreen Memon$^{(\boxtimes)}$, William Liu, and Adnan Al-Anbuky

School of Engineering, Computer and Mathematical Sciences,
Auckland University of Technology, Auckland, New Zealand
`ambreen.memon@aut.ac.nz`

Abstract. The emerging cyber-physical paradigm endeavours to unite all the physical objects embedded with electronics, software, sensors, and network connectivity to allow more direct interactions and information sharing between the physical and cyber worlds. While these massively connected devices and their associated communications can exponentially increase the data generation, transmission, and processing which consume a huge amount of energy and finally end up with harming the environment seriously. In this paper, we propose a solution for energy efficient data dissemination by using the opportunistic device-to-device (D2D) communications. Each sender can decide either use network infrastructure or through encountering the end-users according to the quality of service (QoS) requirements of each data demand and also the mobility behaviors of the users. These decisions are based on the time and location- traces of daily mobility routines and related activities of users and their social relationship. The case study, based on the similarity analysis of the mobility traces, has confirmed the rich opportunities for encountering among people, thus the proposed approach has great promises to reduce the energy consumption of big data dissemination.

Keywords: Big data · Opportunistic routing
Delay tolerant network · Energy efficient data dissemination
Similarity analysis

1 Introduction

The increase in the use of mobile devices has changed the way that users share data and ultimately leading to mobile traffic expansion exponentially since 2011. As per current expectation, more than 24 ex-bytes of mobile traffic will be navigating operators' networks by 2019, with 72% of this traffic being created by the interactive media [1]. In addition, the growing number of mobile devices and their communications have increased significantly the energy consumption.

Delay tolerant networks (DTN) perform store-carry-forward routing to deliver the data in an end-to-end fashion, although a continuous end-to-end

© ICST Institute for Computer Sciences, Social Informatics and Telecommunications Engineering 2018
P. H. J. Chong et al. (Eds.): SmartGIFT 2018, LNICST 245, pp. 155–164, 2018.
https://doi.org/10.1007/978-3-319-94965-9_16

communication path may never exist between sender and destination devices. The integration of the infrastructure-based networks and DTN has shown the benefits since it can boost routing performance and offload traffic from the congested infrastructure networks.

This paper targets to develop a novel eco-friendly and sustainable data transmission approach for offloading the data traffic from the infrastructure to the opportunistic- and social- based device-to-device (D2D) communications. It can be performed by exploring the existing movements and spatial closeness relation among devices. To complement the traditional infrastructure-based data transmission, the new idea is to optimally piggyback data on the moving physical devices for data dissemination to achieve the energy reduction as well as to ensure the QoS requirements. In such way, the proposed approach can fully utilize the users' historical mobility traces to predict the next location. If they are close enough and also satisfy the QoS requirements, the data could be directly transferred using D2D communication which consumes less energy. Otherwise, data could be transferred through the infrastructure-based network.

The rest of the paper is organized as below. Section 2 has reviewed the recent advancements in the areas of D2D communication, delay tolerant network, mobility models and their impact on energy consumption. Section 3 presents the energy consumption model for data dissemination. We introduce the new data dissemination approach by using similarity analysis in Sect. 4. We have conducted a case study in Sect. 5 to validate the new approach and confirm its great promises on substituting the infrastructure-based transmission approach for those delay tolerant data services. Finally, we conclude the main contributions and also layout the future work in Sect. 6.

2   Related Work

In D2D communications, the devices in proximity can interface with each other directly and construct a communication network. Data traffic can be offloaded to the D2D network instead of transmission through the infrastructure based network. For instance, by authorizing D2D communications, some users can download the substances from the cellular base station (BS) while others could get the substances from their associates. Therefore, the D2D communications can significantly reduce the traffic congestions and also energy consumption in networks [2]. In future wireless access networks, balancing the traffic load among base stations can be accomplished through adjusting the user's BS affiliations [3].

At Massachusetts Institute of Technology (MIT), predicting the user behavior is one of their research themes. Time duration is recorded by mobile phones when adjacent to cell tower IDs and Bluetooth devices. While Bluetooth devices demonstrate different behaviors based on how are the devices related to each other. It has been found that the presence of business students in a similar area is performing the same activities [4]. Bluetooth signals were constructed within individual's house to check the accuracy of data transmission into locations by cell towers [5]. Estimating next location of the objectives by using the dynamic

Bayesian Network can reach a successful rate of 93% to 99%. In prediction, the next cell is a sequence of location that they investigated in communication areas, need to improve those resources for reservation and QoS requirements [6]. Zeibart et al. [7] predict that driving to the destinations, given a partially travelled route by calculating the probabilities of different possible routes. Bauer and Deru [8] suggest the ways of predicting future destination along with previous histories. The work in [9] has used the Naive Bayesian classifier based model, which consists of the time slot of days, weekends and 1–8 h. Everyday time stamp is divided into multiple records, that consist of a list of Bluetooth MAC addresses and locations for mobility prediction.

3 Energy Consumption Model

A network can be represented by a graph G (N, L), where N is the number of nodes and L is the direct links (i, j) or edges between graph nodes. For prototyping the key ideas, We assume a general and simplified energy consumption model for wireless or wired energy dissipation where the transmitter dissipates power to generate the radio or line electronics. The power amplifier then consumes energy to transmit the traffic, and the receiver dissipates energy to receive and process the radio or line electronics, as shown in Fig. 1. Taking radio transmission as an instance, the power control can be used to remedy the signal propagation loss by appropriately setting the power amplifier. For example, if the transmission distance is less than a threshold value d_0, the free space prorogation model with the attenuation parameter of ε_{fx} is used, otherwise the multi-path (mp) propagation model with the attenuation parameter of ε_{mp} is used. For the sender to transmit a volume of k-bits data to the receiver where there is a distance of d away, the energy consumption model can be calculated as below:

$$E_{Tx}(k,d) = E_{Tx-elec}(k) + E_{Tx-amp}(k,d) \tag{1}$$

$$E_{Tx}(k,d) = k \cdot E_{elec} + k \cdot \varepsilon_{fx} \cdot d^2, d < d_0 \quad = k \cdot E_{elec} + k \cdot \varepsilon_{mp} \cdot d^4, d > d_0 \tag{2}$$

where, E_{elec} is the energy consumed by the transmitter and it depends on the factors such as digital coding, modulation and filtering signal processing procedures. As for the energy consumed by amplifier, it depends on the distance to the receiver and the acceptable bit-error rate. Energy consumption for the received data can be calculated by:

$$E_{Rx}(k) = E_{(Rx-elec)}(k) = k \cdot E_{elec} \tag{3}$$

Based on the above general energy consumption model for communications, we can see that the volume of data k and the transmission distance d are two critical and changeable factors which can vary the overall energy consumption, compared to the energy consumed by the electronic components and signal processing mechanism in the transmitter, receiver and also the relay amplifiers. It is possible to reduce the transmission distance d which is being traversed through the infrastructure. In other words, the more transmission distance can

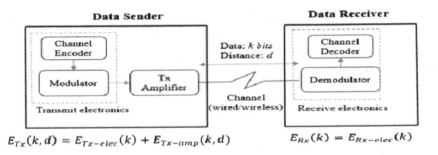

$$E_{Tx}(k,d) = E_{Tx-elec}(k) + E_{Tx-amp}(k,d) \qquad E_{Rx}(k) = E_{Rx-elec}(k)$$

Fig. 1. Energy consumption model

be shortened, the more energy consumption can be reduced. This is the motivation triggering us to propose a new energy efficient data dissemination approach (EEDDA), by fully utilizing the mobility of human (i.e., mobile users) and D2D communications. They carried the data for delivery based on the prediction and users' mobility similarity analysis, especially for those communications services such as file transfer which has delay tolerant characteristic.

4   The Proposed EEDDA Approach

The proposed EEDDA approach, as shown in Fig. 2 has four processes including:

1. Data Collection

   In EEDDA approach, the first step is to collect the users' mobility data. In our case study, we reuse the data gathered by the project of Wireless Topology Discovery (WTD) that was handled at UCSD [10]. It has the traces of 300 people's accessibility of PDAs to WiFi. All the traces has two portions of discussion. One portion consisted of trace data. The other file contained the known locations with access points for local coordinates. Eleven-week trace duration started from the 22 September 2002 to 8 December 2002 was the data collection period.

2. Analysis

   The WTD has sampled and recorded the above information for all access points (APs) for every 40 s, which may fill in all its frequencies. The analysis of the collected data was conducted while running on a student's device. During a sample, the above information is given by WTD for all sensed samples. In Fig. 3 below, the three entries were recorded if a device has three APs in one sample (the entries include the IP address, signal strength, and attached flag). To extract basic records that show the user's location, the Associate field is used in this study. The user's device is located near the connected access point (Associated = 1) is based on the assumption. A list of neighboring access points is created for the sensitive access points that were not selected for the Association (Associated = 0). To record that a user should be decided at any time, AC-POWER field can be used for it. The assumption is that

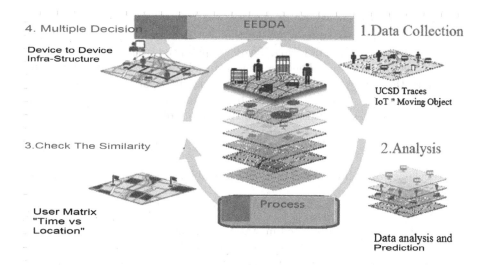

Fig. 2. The Proposed EEDDA approach

USER_ID	SAMPLE_DATE & TIME		AP_ID	SIG_STRENGTH	AC_POWER	ASSOCIATED
123	Sep-22	0:00:00	359	8	0	0
123	Sep-22	0:00:00	363	5	0	0
123	Sep-22	0:00:00	365	11	0	1

Fig. 3. The fields and usage of the database

the user is not mobile and AC-POWER = 1 is a plug-in in the device. In weight gain of individual access points, the given time, signaling power SIG-STRENGTH can be used.

In different location alignment algorithms, the SIG-STRENGTH field can be used. For example, a user is using the trilateration algorithm then it is between all access points. The AP-ID uses only one assigned number to label every access point in the field databases. Here we use these values as a location label, and when evaluating future locations, these values are developed by the model. In a real application, AP-ID field will be backed up on the map or on a map named on a useful location. The SAMPLE-TIME contains the date and time. The recorded state of every 20 s and end of 11 weeks period by their own devices. Throughout the pre-processing, the week fragments at the start and end of during the 11 weeks were rejected so as to provide 10 whole weeks' samples with User ID field for each record to the specific user. This algorithm does not use the user information, but simply uses the user-id field for partitioning the logs into individual user logs. The prediction can be done by individual user but not on the entire group.

3. Check Similarity

The sequence prediction is to predict the next item in a sequence, which can be considered as a kind of rating. There are potential results for the alphabet used to create a layout. They are known in advance and predict the model in which the next item is in sequence. First signify theorizing symbols S_1, S_2, S_3, S_4, S_n. The n represents the number of symbols in the alphabet. When training sequences are described with the symbol of t:

$$Where\ X_i \in \sum X_1, X_2, X_3...X_t \qquad (4)$$

this calculation defines the conditional probability

$$Pr\left\{X_{t+1}x_{t+1}|X_t = x_t, x_{t-1} = x_{t-1}\right\} \qquad (5)$$

This calculation has been used in stationary Markov chain [11]. In our case, we are considering the stationary one, because the probabilities are not only depend on the same time, even sub-sequences repeat at the same time but with different location in sequences for each repeat or shift S and for all X_i.

$$Pr\left\{X_1, X_2 = x_2.....x_nX_1, X_2 = x_2.....x_n\right\}$$
$$Pr\left\{X_{1+m} = X_1, X_{2+M} = x_2.., X_{n+m} = x_n\right\} \qquad (6)$$

This process is called Markov model because the probability is likely to be on the variable. The number used is the variable for variants, L, model length, or order. The pre-variable sub-division is called history or context. If the contextual length of the context is set continuous, the model is called fixed length Markov chain [12]. The variable length Markov chain with length L, the context length used to vary the maximum number of L on the prediction of the Markov channel. The first order of Markov model makes the basis for prediction model here.

Moreover, the raw data from the WTD experiment combines all logs, from all operators into one file. The entire logs of all the operators are associated with the document by the raw data from the WTD test. The first step of associating all logs from different users into one folder was to divide the raw information into various documentation for an individual user. Records usually here refer to as sensed, non-associated, access points. Records with the same date and time had the maximum amount of indication matrix. Whereas the rest records were rejected. Reports with similar user name, access point and having a starting time within one minute of the preceding record's starting time are termed as contiguous records. The output showed the length of each session collected over a day. The statistics recorded by a full-time active mobile device is regarded as the first movement data. The data representing users' significant locations revealing social interacting applications is called the destination data. In this work, significant locations are determined solely by a length of stay of at least 10 min [12]. The destination data and movement could be considered as movement location and significant location

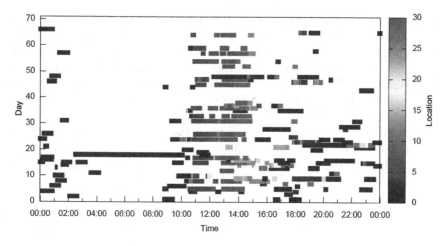

Fig. 4. Data (1-minute or less) for User 003 (Color figure online)

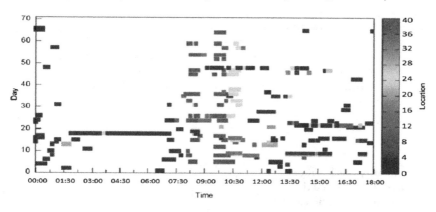

Fig. 5. Data (10-minute) for User 003

individually. The starting time is measured to be on entire number nearest to the last minute dataset of MoveLoc. Later on, all the duration include those, slightly lesser than 20 s, were included of one minute window. When the session started then it was considered as the highest period of one-minute time duration, declining the rest our work prediction on the future location and time. The Fig. 4 shows the MoveLoc data for user 3. The color bar shows that this user visited 40 locations, over the 10-weeks recording period. The x-axis represents the time of day. Anyone can observe that this user had some regular locations between 1:00 pm and 2:00 pm and some of the movement between locations is indicated by the change in colors/shades. Figure 5 depicts that User 3 MoveLoc (1-minute or less) and User 3 SigLoc (10- minute) data and dataset location. Where user consumed the minimum of 10 min, were covered in the SigLoc dataset. Elimination other sessions less than 10 min was the first phase of producing this assessment. The start-time of the other sessions were almost equalized to the bordering 10 min. The SigLoc data for User 3 is indicated in Fig. 4. One can notice that the number of locations fell from 40 to 20 and the transitions between locations were removed. Though the existing point of proof-of-concept on EEDDA, which does not advance towards a far-reaching position, yet the study outcome is relatively stimulating.

4. Multiple Decision

This is the last process of EEDDA, of which the device will be able to take the decision as per the result of the similarity analysis. The data is either transmitted through the infrastructure based network or D2D communications. The Fig. 6 shows the flow chart of the decision process.

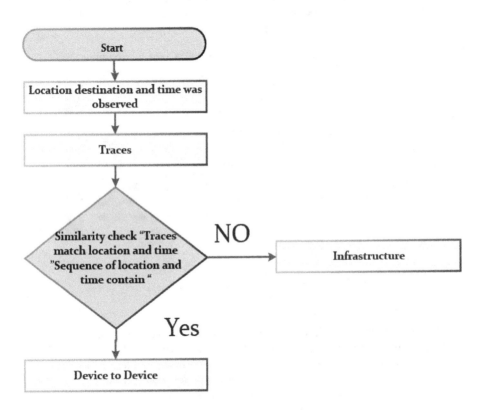

Fig. 6. The process of multiple decision

5   A Case Study

We have identified that the data transmission distance is a key but changeable factor contributing to the overall energy consumption. Here we consider a scenario, of which a professor in university and his students have communicated with each other very frequently. They are all in the same building but on the different floors. We assume that the data is delay tolerantable. In such way, we have two options including option A: send the data through infrastructure-based network, and option B: infrastructure-less approach i.e., D2D communications.

To disseminate data through these options, we assume further two possible ways: (a) Human mobility traces get matched according to the movement behavior. In such way, the proposed approach can fully utilize the benefits of mobility traces and check similarity of location and time. When they are matched, the data can transfer through D2D while consuming less energy. (b) In this option, we use the prediction model with human mobility traces that checks predicted location and time of user's movement. When it matches, the data will transfer by D2D communications.

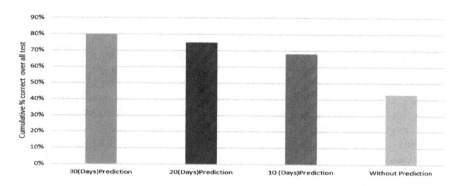

Fig. 7. Test result with 30 days, 20 days, 10 days interval prediction and without prediction

Figure 7 shows the correctness of the results as per time allocated for prediction. If we simply use the traces, the accuracy of the results is 45%, while considering delay for about 10 days interval. While in the prediction option, it gets increased by 70%. However, when we increase the interval of delay for about 20 days or 30 days, the accuracy could be better than the previous. This indicates that the longer tolerated delay and mobility duration, the higher opportunity that two users encounter each other to have D2D communications, which could reduce the energy consumption significantly.

6   Conclusion and Future Work

In this paper, a novel energy efficient data dissemination approach (EEDDA) is introduced, of which a mobile encounter between the communicating pairs is sought for directly exchanging data. The similarity analysis framework reveals the mobile encounter opportunities among communication peers. Peer device similarity depends on the moving ability and adaptability of interacting with the devices. The work is ongoing for developing communication protocol and algorithm but also for peer countering and automation. The greater number of complex mobility models and the unpredictability of devices' movements should be considered too to develop more adaptable EEDDA in future work.

References

1. Micinski, K.: Interaction-based security for mobile apps. Ph.D. dissertation, University of Maryland, College Park (2017)
2. Kjærgaard, M.B., Langdal, J., Godsk, T., Toftkjær, T.: Entracked: energy-efficient robust position tracking for mobile devices. In: Proceedings of the 7th International Conference on Mobile Systems, Applications, and Services, pp. 221–234. ACM (2009)
3. Dogar, F.R., Steenkiste, P., Papagiannaki, K.: Catnap: exploiting high bandwidth wireless interfaces to save energy for mobile devices. In: Proceedings of the 8th International Conference on Mobile Systems, Applications, and Services, pp. 107–122. ACM (2010)
4. Ziebart, B.D., Maas, A.L., Dey, A.K., Bagnell, J.A.: Navigate like a cabbie: probabilistic reasoning from observed context-aware behavior. In: Proceedings of the 10th International Conference on Ubiquitous Computing, pp. 322–331. ACM (2008)
5. Elnekave, S., Last, M., Maimon, O.: Predicting future locations using clusters' centroids. In: Proceedings of the 15th Annual ACM International Symposium on Advances in Geographic Information Systems, p. 55. ACM (2007)
6. Roddick, J.F., Spiliopoulou, M.: A survey of temporal knowledge discovery paradigms and methods. IEEE Trans. Knowl. Data Eng. 14(4), 750–767 (2002)
7. Mayrhofer, R., Radi, H., Ferscha, A.: Recognizing and predicting context by learning from user behavior. na (2003)
8. Holbach, H., Bourassa, M.A.: The effects of gap wind induced vorticity, the ITCZ, and monsoon trough on tropical cyclogenesis. In: 2012 International Ocean Vector Winds Science Team Meeting (2012)
9. Bellotti, V., Begole, B., Chi, E.H., Ducheneaut, N., Fang, J., Isaacs, E., King, T., Newman, M.W., Partridge, K., Price, B., et al.: Activity-based serendipitous recommendations with the magitti mobile leisure guide. In: Proceedings of the SIGCHI Conference on Human Factors in Computing Systems, pp. 1157–1166. ACM (2008)
10. UCSD WTD Trace. http://sysnet.ucsd.edu/wtd/
11. Yang, G.: Discovering significant places from mobile phones – a mass market solution. In: Fuller, R., Koutsoukos, X.D. (eds.) MELT 2009. LNCS, vol. 5801, pp. 34–49. Springer, Heidelberg (2009). https://doi.org/10.1007/978-3-642-04385-7_3
12. Vu, L., Do, Q., Nahrstedt, K.: Exploiting joint wifi/bluetooth trace to predict people movement (2010)

Prediction of Electricity Consumption for Residential Houses in New Zealand

Aziz Ahmad[1], Timothy N. Anderson[2], and Saeed Ur Rehman[2(✉)]

[1] Unitec Institute of Technology, Auckland 1142, New Zealand
aahmad@unitec.ac.nz
[2] Auckland University of Technology, Auckland 1010, New Zealand
saeed.rehman@aut.ac.nz

Abstract. Residential consumer's demand of electricity is continuously growing, which leads to high greenhouse gas emissions. Detailed analysis of electricity consumption characteristics for residential buildings is needed to improve efficiency, availability and to plan in advance for periods of high electricity demand. In this research work, we have proposed an artificial neural network based model, which predicts the energy consumption of a residential house in Auckland 24 h in advance with more accuracy than the benchmark persistence approach. The effects of five weather variables on energy consumption was analyzed. Further, the model was experimented with three different training algorithms, the levenberg-marquadt (LM), bayesian regularization and scaled conjugate gradient and their effect on prediction accuracy was analyzed.

Keywords: Electricity demand prediction · Load prediction · Neural network
Load management

1 Introduction

The residential sector of a country consumes 16–50% of energy of all sectors and averages approximately at 30% globally as shown in Fig. 1. The relatively high energy consumption of residential users demands a detail analysis of its energy consumption characteristics in order to promote conservation, efficiency, technological implementation and energy source switching, such as the utilization of on-site renewable energy sources [1].

Recent increase in the implementation of renewable energy systems has increased interest in the precise modeling and prediction of energy consumption. Energy production from renewable sources vary throughout 24 h 365 days a year. Therefore, predicting the energy consumption 24 h prior helps in efficient optimization of energy distribution among loads specifically between building and local grids. On the other hand, electricity consumption prediction is essential for generators, wholesalers and retailers of electric energy, who buy and sell, switch loads, plan maintenance and unit commitment and much more. However, with increasing costs passed to consumers, the optimum on-site energy generators can be achieved only when consumers can forecast their requirements in order to efficiently utilize grid-tied storage system and solar panels, thus enabling intelligent buildings.

© ICST Institute for Computer Sciences, Social Informatics and Telecommunications Engineering 2018
P. H. J. Chong et al. (Eds.): SmartGIFT 2018, LNICST 245, pp. 165–172, 2018.
https://doi.org/10.1007/978-3-319-94965-9_17

In [2], a building that minimizes the grid power usage and maximizes services efficiency is defined as an intelligent building. Sensors, renewable energy sources and energy management system (EMS) are some of the components of intelligent building with high ranking given to EMS, which monitors and controls the energy consumption. However, the effective operation of an EMS can be realized by precisely predicting the required power consumption. The early predication of load consumption would enable the planning of load distribution in advance and avoid power outages by shifting the unnecessary load to off peak time, to reduce cost and increase efficiency.

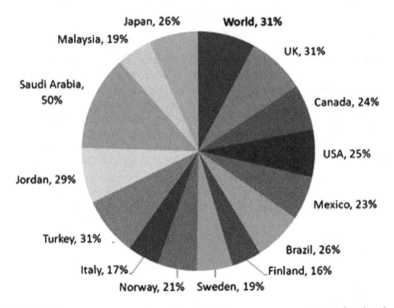

Fig. 1. Worldwide energy consumption of residential users as a percentage of national energy usage of different sectors [1].

Short term load prediction of a residential house is a complex task due to the usage of various equipment with varying power requirement. Recently, short term load prediction has attracted considerable attention from researcher and scientists working both in academia and industry. A number of mathematical models and machine learning algorithms have been investigated for short term load prediction both in residential and small industry. Some of the algorithms used are regressive analysis, wavelet analysis, fuzzy system modeling, neural network modeling, evolutionary algorithms and hybrid approaches [3–10].

Artificial neural network (ANN) has been investigated for load forecasting mostly in commercial domain for large loads, such as ANN expert system [11–13]. ANN has the advantage to implicitly extract non-linear relationship among loading variables, learn and predict the future values. ANN has applications in different areas of power systems to handle complex non-linear functions, provide reliability and efficiency even for the cases where learning data is incomplete or not available at all [13, 14]. In majority of previous work, ANN has been used for large scale forecasting. This research work uses different training algorithms of ANN for short-term load forecasting for a residential house.

Our main contribution of this research work is to train the ANN model with three different algorithms to predict electric load in a typical New Zealand residential house. A typical residential consumer would be living in a three bedroom house with four occupants (2 adults and 2 children). The goal is to predict the load 24 h in advance, so that renewable energy sources and its components can be efficiently utilized. It is suggested that a similar model could be adopted for other locations with varying number of occupants.

2 Methodology

In this research work, an ANN based NARX (nonlinear autoregressive network with exogenous inputs) predictive model was used to forecast future values of electricity consumption. The NARX model uses historical values of electricity consumption and historical values of five environmental variables one of the input variable is hourly electricity data collected from a residential house in Auckland, New Zealand. The five environmental variables are: Temperature (T_{mean}), Barometric Pressure (P), Relative Humidity (RH), Wind speed (W_s) and Wind direction (W_d). Data were taken from the National Institute of Water and Atmosphere's (NIWA) CliFlo database (2014) [17]. Our ANN model was trained on the historical electricity consumption and environmental variables data with electricity consumption 24 h in advance being the output variable.

The predictive model can be expressed mathematically by predicting future values of the electricity consumption time series y(t) from past values of that time series and past values of input variables time series x(t) [15]. The equation for the NARX model is given by Eq. 1.

$$y(t) = f\big(y(t-1), y(t-2), \ldots, y(t-n_y), u(t-1), u(t-2), \ldots, u(t-n_u)\big) \qquad (1)$$

Where the next value of the dependent output signal y(t) is regressed on previous values of the output signal and previous values of an independent input signal. The NARX model is implemented using a feed-forward neural network to approximate the function f in Eq. 1. Figure 2 shows the diagram of the resulting network, where y(t) output series is predicted given past values of y(t) and another input series x(t).

The prediction accuracy of ANN models is dependent on the combination of weather predictor variables and training algorithm [16].

Various combinations of weather predictor variables were tested using a statistical analysis to find the relationship between electricity consumption and individual weather variable as shown in Table 1. The Pearson correlation coefficient (R) is a measure of the linear correlation between two variables, giving a value between +1 and −1 inclusive, where 1 is total positive correlation, 0 is no correlation, and −1 is total negative correlation.

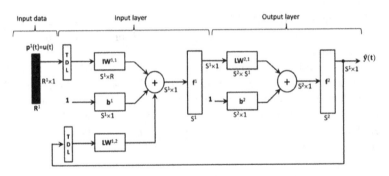

Fig. 2. NARX network diagram

In order to determine the performance of the ANN models, the regression (R) and mean squared error (MSE) values were analyzed. The mean squared error (Eq. 2) provides information on the short term performance and is a measure of the variation of predicated values around the measured data, lower MSE values represent better prediction accuracy.

$$MSE = \frac{1}{N}\sum_{i=1}^{N}(I_{p,i} - I_i)^2 \tag{2}$$

Where the predicted electricity consumption in kWh is I_p, I_i is the measured electricity consumption in kWh, and N denotes the number of observations.

Table 1. Regression (R) values for input weather variables vs electricity consumption

Input variables	R
Temperature (T_{mean})	0.438
Relative Humidity (RH)	0.335
Pressure (P)	0.203
Wind speed (Ws)	0.033
Wind direction (Wd)	0.030

Further, the models were experimented with three different training algorithms, the levenberg-marquadt, bayesian regularization and scaled conjugate gradient and their effect on prediction accuracy was analyzed as shown in Table 2.

Table 2. Effect of training algorithms on prediction accuracy

Input variables	MSE (kWh)	R	Processing time
Levenberg-Marquadt	0.125	0.438	00:00:01
Bayesian regularization	0.129	0.417	00:01:53
Scaled conjugate gradient	0.142	0.360	00:00:02

2.1 Electricity Consumption Prediction for a Residential House

Figures 3 and 4 show four-days and one-day prediction of electricity consumption for a residential house in Auckland with 2 adults and two children respectively. It can be seen in Figs. 3, 4 and Table 2 that the LM training algorithms predicts electricity consumption more accurately than the BR and SCG training algorithms.

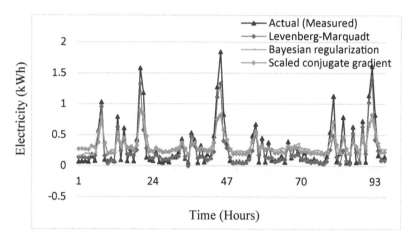

Fig. 3. The predicted and measured values of electricity consumption for four-days with MSE = 0.125 kWh

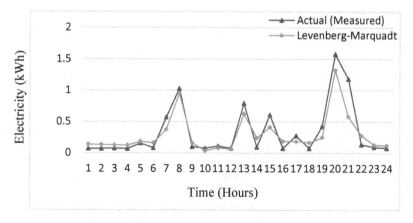

Fig. 4. 24-h measured and predicted values of electricity consumption for the LM algorithm

2.2 A Benchmark Persistence Model

As a comparative study, the persistence model was developed using Eq. 3 to predict the h hour-ahead forecasting (h = 1, 2, 3 … hours).

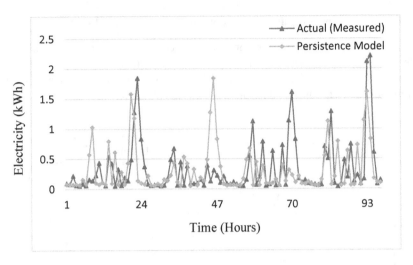

Fig. 5. Four-days measured and predicted values of electricity consumption using the benchmark persistence approach with MSE = 0.998 kWh

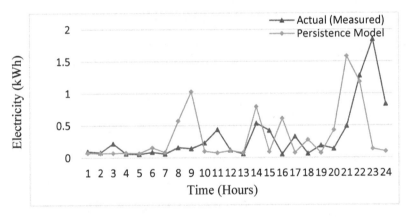

Fig. 6. 24-h measured and predicted values of electricity consumption using the benchmark persistence approach with MSE = 0.998 kWh

$$S(t+h) = S(t) \tag{3}$$

Where $S(t+h)$ is the predicted electricity consumption at time $t+h$.

Same electricity consumption and five input variables data used for the NARX model was utilized for the benchmark persistence approach, with electricity consumption as the objective function. Mean square error as defined in Eq. 2 was calculated to validate and compare the model performance. Figures 5 and 6 show four-days and one-day prediction of electricity consumption for a residential house in Auckland with 2 adults and two children respectively.

3 Conclusion

In this study, the effect of five weather variables on electricity consumption was studied using the linear regression analysis and observed that temperature and relative humidity showed correlation with the electricity consumption with regression values of 0.438 and 0.335 respectively. A predictive model based on the recurrent neural network, consisting of mean temperature as input and electricity consumption as the target variable was tested with three training algorithms, the levenberg-marquadt (LM), bayesian regularization and scaled conjugate gradient. LM back propagation algorithm produced the lowest mean square error of 0.125 as compared to 0.129 and 0.142 for the bayesian regularization and scaled conjugate gradient algorithms respectively. The proposed model was also compared with a benchmark persistence model and it was found that the proposed approach predicted electricity consumption values 24-h ahead with more accuracy than the persistence approach for a house in Auckland with two adults and two children.

References

1. Lukas, G.S., Ugursal, V.I.: Modeling of end-use energy consumption in the residential sector: a review of modelling techniques. Renew. Sustain. Energy Rev. **13**, 1819–1835 (2009)
2. Flax, B.: Intelligent buildings. IEEE Commun. Mag. **29**, 24–27 (1991)
3. Xia, C., Wang, J., McMenemy, K.: Short, medium and long term load forecasting model and virtual load forecaster based on radial basis function neural networks. Int. J. Electr. Power Energy Syst. **32**, 743–750 (2010)
4. Maia, C., Goncalves, M.: A methodology for short-term electric load forecasting based on specialized recursive digital filters. Comput. Ind. Eng. **57**, 724–731 (2009)
5. Niu, D., Gu, Z., Zhang, Y.: An AFSA-TSGM based wavelet neural network for power load forecasting. In: Yu, W., He, H., Zhang, N. (eds.) ISNN 2009. LNCS, vol. 5553, pp. 1034–1043. Springer, Heidelberg (2009). https://doi.org/10.1007/978-3-642-01513-7_114
6. Tso, G.K.F., Yau, K.K.W.: A study of domestic energy usage pattern in Hong Kong. Energy **28**, 1671–1682 (2003)
7. Al-Garni, A.Z., Zubair, S.M., Nizami, J.S.: A regression model for electric energy consumption forecasting in Eastern Saudi Arabia. Energy **19**, 1043–1049 (1994)
8. Yan, Y.Y.: Climate and residential electricity consumption in Hong Kong. Energy **23**, 17–20 (1998)
9. Ranjan, M., Jain, V.K.: Modelling of electrical energy consumption in Delhi. Energy **24**, 351–361 (1999)
10. Egelioglu, F., Mohamada, A.A., Guven, H.: Economic variables and electricity consumption in Northern Cyprus. Energy **26**, 355–362 (2001)
11. Ho, L.K., Hsu, Y.Y., Chen, F.C., Lee, E.T., Liang, C.C., Lai, S.T., Chen, K.K.: Short term load forecasting of Taiwan power system using a knowledge-based expert system. IEEE Trans. Power Syst. **5**, 1214–1221 (1990)
12. Rahman, S., Hazim, O.: A generalized knowledge-based short-term load-forecasting technique. IEEE Trans. Power Syst. **8**, 508–514 (1993)
13. Moghram, I., Rahman, S.: Analysis and evaluation of five short-term load forecasting techniques. IEEE Trans. Power Syst. **4**, 1484–1491 (1989)

14. Khotanzad, A., Afhkhami-Rohani, R., Maratukulam, D.: ANNSTLF artificial neural network short-term load forecaster generation three. IEEE Trans. Neural Netw. **13**, 1413–1422 (1998)
15. Eugen, D.: The use of NARX neural networks to predict chaotic time series. WSEAS Trans. Comput. Res. **3**, 182–191 (2008)
16. Yadav, A.K., Chandel, S.S.: Solar radiation prediction using artificial neural network techniques: a review. Renew. Sustain. Energy Rev. **33**, 772–781 (2014)
17. National Institute of Water and Atmospheric Research. http://cliflo.niwa.co.nz/. Accessed 10 Oct 2018

Physical-Layer Network Coding
with High-Order Modulations

Xuesong Wang[1,2,3] and Lu Lu[1,2(✉)]

[1] Key Laboratory of Space Utilization,
Technology and Engineering Center for Space Utilization,
Chinese Academy of Sciences, Beijing, China
lulu@csu.ac.cn
[2] University of Chinese Academy of Sciences, Beijing, China
[3] Science and Technology on Communication Networks Laboratory,
Shijiazhuang, China

Abstract. Physical-Layer Network Coding (PNC) can double the throughput of a Two-Way Relay Network (TWRN) by reducing packet exchanging timeslots. In a multi-user wireless communication system, time domain phase shift can inevitably lead to deterioration of PNC performance. In previous studies, there have been many studies result to enhance the performance of some low-order modulation techniques such as BPSK and QPSK, but fewer studies are designed for high-order modulation such as 16-QAM. It is known that high-order modulation is the only way to improve the spectrum utilization rate. This paper uses simulation to explain that the time domain phase shift will greatly affect the performance of 16-QAM PNC, and its' performance couldn't be improved even polar code is used. To address this phase penalty problem, we propose a half-symbol asynchronous algorithm to introduce correlations using belief propagation (BP). Simulation results show that the time domain phase shift problem of 16-QAM modulated PNC systems can be solved effectively using our proposed half-symbol asynchronous BP algorithm.

Keywords: Physical-Layer Network Coding · Phase asynchrony
Symbol asynchrony · High-order modulation · Belief propagation

1 Introduction

The concept of Physical-Layer Network Coding (PNC) was first proposed in 2006 [1, 5]. PNC could increase throughput by reducing transmission slots. In wireless communications, various electromagnetic signals transmitted in space are superimposed within the channel, but only the signal sent by the specific terminal is useful to

The work of L. Lu was supported by the NSFC under Project 61501390. The work of X. Wang was supported in party by the Science and Technology on Communication Networks Laboratory under Project 614210401050217, and in part by the Special Presidential Foundation of Technology and Engineering Center for Space Utilization of the Chinese Academy of Sciences under Project CSU-ZDBS-201702.

the communication nodes, while other signals are considered as interferences. This interference has brought the obvious consequence to the multi-hop point-to-point network. For example, in the 802.11 protocol network, the throughput of a single hop network can reach 4 times which multi-hop network does theoretically [2]. One common solution is to use relay nodes to forward signals from other sources. The other solution is to specify a fixed communication protocol to prevent multiple nodes in the same channel sending messages at the same time. The latter way will result in a significant increase of the total transmission time slots in the whole communication process.

The PNC allows the sender to send information simultaneously to the relay, exploiting this "jamming" by implementing special information processing at the relay without affecting the reliability of the communication. As a result, PNC received widespread concern from the communications community.

Related Works. Since the PNC concept was proposed, a variety of PNC-based applications are also emerging. Zhang et al. discussed the feasibility of PNC in Galois Fields. Liew, Lu and Zhang performed a great deal of work on PNC, including channel coding PNC, asynchronous PNC, FPNC with OFDM and so on, and conducted a large number of experiments on Universal Software Radio Peripheral (USRP) [1, 3, 8]. Pan et al. conducted a study of 16-QAM with PNC combined with multiple access and MIMO techniques, but the solution will increase the cost by generating redundant information during demodulation, and it lacks the possibility of extending to higher order modulation [6].

With the application of channel coding, PNC can improve its code error correction capability to ensure the reliability of transmission. L. Chen combined Lattice coding with PNC to prove that Lattice physical layer network coding has superior performance, but the main result is hard to achieve due to its complexity [4]. Du et al. proposed a PNC scheme using LDPC under Gaussian Two-Way Relay Network (TWRN), and experimentally verified that it has an improvement on the Bit Error Rate (BER) at higher-order PSK [7]; however, QAM is more reliable than PSK at higher orders and there is a lack of QAM studies in it.

Due to the limited spectrum resources, high-order modulation is needed to improve the utilization of spectrum resources. However, the studies above are either based on simple modulation schemes such as BPSK, QPSK, or using high order modulation but lack of scalability. Moreover, many papers assume that the signal received by the relay has no phase offset. Even if there is phase offset, the study mainly focuses on simple modulation and does not involve high-order modulation. Therefore, this paper will mainly study the performance problems of PNC under high-order modulation, and a scheme of channel coding 16-QAM PNC will be introduce below. In addition, in this paper, and a scheme using half-symbol misalignment with belief propagation (BP), which could solve the performance problems, will be shown in detail. And this solution has the ability of extending to higher order modulation.

In summary, the contributions of this paper are as follows.

- We rigorously study a 16-QAM modulated PNC system in a TWRN under regular PNC mapping rules. We find that phase asynchrony can lead to tremendous system BER performance loss.

- We put forth a belief prorogation based decoding algorithm, by purposely introducing a half-symbol asynchrony to the 16-QAM modulated PNC system, to address the phase penalty issue.
- System performances of both non-channel-coded and channel-coded 16-QAM PNC system are discussed.

The following sections are organized as follows. Section 2 presents the PNC system model with 16-QAM modulation. Section 3 points out the phase asynchrony problem in time-domain PNC systems. Section 4 studies and observes the performance of 16-QAM PNC with polar coding. Section 5 puts forward the solution to the time-domain phase shift and observes the results. Section 6 concludes the paper.

2 16-QAM PNC System Model

2.1 PNC Transmission Model

PNC is mainly used in Two-Way Relay Network (TWRN), which is a three-node communication model. In TWRN, node 1 and 2 transmit information to each other through relay node R. A real example could be that two base stations which are very far apart want to communicate via satellites. Assuming it is a half-duplex situation, in this case any node cannot send or receive data at the same time. In traditional scheme, the two communication nodes send messages to the node R in different timeslot to avoid conflict with each other, which needs four timeslots in total. The process of PNC transmission mode is shown in Fig. 1. In the uplink, two communication nodes 1 and 2 transmit message and at the same time. After the relay R receives this superposed "naturally generated" sum information $(M_1 + M_2)$, it uses the corresponding mapping rule to coding the message

$$M_R = g\ (M_1 + M_2) = M_1 \oplus M_2 \tag{1}$$

then the M_R will be sent back to both source nodes at the same time. In the downlink, the two source nodes receive the message and then recover the message sent by other side with the copy of the local message. With the "interference" being used here, the system will no longer limit whether the communication node can send message at the same time. The total use of the slot reduced to two, thereby the system throughput is increased in the same period of time.

We can see that PNC uses the expense of increasing local copy and increasing the processing overhead at the relay to exchange for a reduction of the time slot and an increase in throughput. However, this burden is negligible comparing to the benefit here, which is, the "interference" caused by the simultaneous transmission of information by nodes 1 and 2 becomes a part of the network coding calculation, thus the influence of interference is eliminated.

As can be seen from the process above, one of the key issues of PNC is how to achieve Eq. (1), which is, how to complete the mapping from $M_1 + M_2$ to $M_1 \oplus M_2$. This problem will be described below.

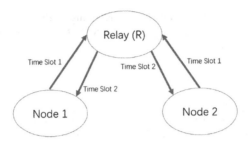

Fig. 1. PNC transmission mode

2.2   16-QAM PNC Mapping Rules

Here we establish the mapping rule under 16-QAM PNC system. In fact, any method who accomplishes a one-by-one mapping rule from to is available; but in this paper, a bit-by-bit mapping rule is established because it is one-to-one mapping on amplitude, and the nodes are easy to make a judgement.

When discussing "bit-by-bit mapping", it has to be mentioned that Gray coding is used in 16-QAM in this paper, since it is a kind of error minimization coding method because there is only one bit difference between the adjacent constellation points

The general model representation of 16-QAM PNC is consistent with the PNC model of Sect. 2.1, but it's more complicated comparing with BPSK or QPSK. 16-QAM involves I and Q components in different amplitudes. Section 2.1 states that the key issue with 16-QAM is the mapping of $M_1 + M_2$ to $M_1 \oplus M_2$. When discussing the mapping rule, only the mathematical realization is considered; which is, the factors in the actual communication is irrespective here. For 16-QAM, the mapping can be divided into I components and Q components separately, and the two mappings are the same.

When node 1 or 2 performs 16-QAM modulation and sends information, its I and Q components correspond to the mapping relationships in Table 1, where m is a bit sequence, s is the operator performing an eXclusive OR operation (XOR), and a is amplitude. Table 1 applies for any 16-QAM I/Q component involved in the overall system.

Let node 1 send the message M_1 with the expression

$$x_1 = a_1 \cos \theta + b_1 \sin \theta \tag{2}$$

and let node 2 send the message M_2 with the expression

$$x_2 = a_2 \cos \theta + b_2 \sin \theta \tag{3}$$

then the relay will receive the message M_R with the expression

$$x_R = (a_1 + a_2) \cos \theta + (b_1 + b_2) \sin \theta = a_R \cos \theta + b_R \sin \theta \tag{4}$$

where $a_1, a_2 \in \{-3, -1, 1, 3\}$, $a_R \in \{-6, -4, -2, 0, 2, 4, 6\}$.

Table 1. Correspondence of one component in 16-QAM.

Bit sequence m^I/m^Q	XOR operator s	amplitude a
00	0	-3
01	1	-1
11	2	1
10	3	3

After receiving x_R, the relay obtains the 16-QAM massage M_R through the network coding mapping function with the expression Eq. (1) and then sends it out. The I/Q components of the superposition signal y_R have seven kinds of amplitude, but components in M_R are still four kinds of amplitude. Here is an example using I component to specify the mapping process. The method is the same with Q component.

Let $m_1^I/s_1/a_1$ represent the bit sequence, the operator of the XOR operation, and the modulation amplitude of the I component of the node 1, respectively. Let $m_2^I/s_2/a_2$ represent the bit, the operator of the XOR operation, and the modulation amplitude of the I component of the node 2, respectively. If node 1 send the bit 1110, which means $m_1^I = 11$; node 2 send the bit 1001, which means $m_2^I = 10$, then according to the Table 1, the corresponding XOR operator should be $s_1 = 2$ and $s_2 = 3$. So, the relay does XOR calculation to get the operator

$$s_R = (s_1 + s_2) \bmod 4 = (2+3) \bmod 4 = 1 \tag{5}$$

As we can see that in Table 1, s_R is corresponded to bit $m_R^I = 01$ (when performing XOR by bit, we can get $m_1^I \oplus m_2^I = 11 \oplus 10 = 01 = m_R^I$), and the corresponding signal amplitude is $a_R = -1$. The signal amplitude previously received by R is $a_1 + a_2 = 1 + 3 = 4$. That is, if the amplitude of the component in the signal that the relay receives is $a_1 + a_2 = 4$, then it should be mapped $a_R = -1$ as its amplitude. The relay may not care about the value of s_R or m_R^I. The information contained can be given to node 1 or 2 to deal with. The relay's only mission is to get the correct transmission amplitude on it. Suppose that the relay sends a signal with one of the component is $a_R = -1$, and it is received by node 1. Node 1 maps it to $m_R^I = 01$ and matches the local copy bit m_R^I to perform XOR

$$m_1^I \oplus m_R^I = 11 \oplus 01 = 10 = m_2^I \tag{6}$$

then the message sent by node 2 is known by node 1.

When discussing the amplitude mapping method at the relay R, there are 16 cases for the component mapping at the relay R, since the two 16-QAM symbols' I and Q components have 4 amplitudes each [3]. Note that other corresponding solutions are also available. This article creates this mapping method because it is all one-to-one mapping on amplitude; using other methods may result in one-to-many mapping at some points, which in some cases judgment cannot be done. It can be obtained that the relay at the signal processing need to do the key mapping shown in Table 2.

Table 2. Key mapping at the relay.

$a_1 + a_2$	-6	-4	-2	0	2	4	6
a_R	-3	-1	1	3	-3	-1	1

Table 2 shows the key mapping, which means $a_1 + a_2$ to a_R. The previous example shows that the relay does not care about the value of s_R or the value of m_R, as long as the amplitude is mapped correctly. In this way, the mapping rule required under 16-QAM is obtained.

2.3 Establishment of 16-QAM PNC Model

This section describes the general mathematic models and the formulations of 16-QAM PNC systems involved in this dissertation for the remainder of the paper. There are many practical problems to consider about PNC, such as phase offset in time domain, phase offset in frequency domain, noise problem, channel coding problem, symbol synchronization problem, channel fading problem, and so on. Since the problem of phase shift in frequency domain can be solved by using Orthogonal Frequency Division Multiplexing (OFDM) [8], the problem of phase shift in time domain is mainly addressed in this dissertation. The phrase "phase offset" in this paper below refers to the time domain phase offset. Symbol synchronization is assumed here; in Sect. 5, symbolic asynchrony exists as a necessary condition, so the problem with asynchronous systems will be explained later. Channel fading is not a concern here. Additive White Gaussian Noise (AWGN) runs through the entire study. Section 4 will focus on the problem of 16-QAM PNC channel coding; for simplicity, channel coding is not considered here. Since the situation of PNC downlink is similar to that of Point to Point (P2P), this paper mainly considers the PNC uplink.

In general, the phase offset in the time domain is due to the asynchronous phase of the carrier frequency oscillator or to the different path delays of the two uplinks (node 1 to node R, node 2 to node R). The time-domain phase shift will cause the constellation points to twist when demodulating, resulting in the failure of demodulation and the increase of BER [10]. Assuming that two uplinks are symmetrical and constant-parameters channels, the phase offset due to the differences in path delays may not be considered. Then it can be assumed that the phase offset is mainly due to the frequency difference of the local oscillator of each node. Then, the relay receives the signal from node 1 and 2 to estimate the frequency difference in coherent demodulation between R and 1 (φ_1) or R and 2 (φ_2). Without loss of generality, assuming $\varphi_2 > \varphi_1$, then

$$\varphi = |\varphi_1 - \varphi_2| = \varphi_2 - \varphi_1 \tag{7}$$

is the frequency difference between the two uplinks obtained by the relay, which is the phase offset to be studied in this paper. It is assumed that specific value of φ can be estimated by relay, which can be solved by technical methods today. In this way, this article can focus on the impact of φ.

For a signal $x(t)$, it is mathematically possible to know $x(t) = x(t)e^{2N\pi}, N \in \mathbf{Z}$, which means the phase shift varies over each 2π period. For simplicity, this article mainly studies the situation of $\varphi \in [0, \pi]$.

Since this paper only considers the phase offset φ in the time domain and the noise n in AWGN channel, it can be assumed that a certain symbol with its length N sent by node 1 is $m_1[n](n = 1, 2, \ldots, i, \ldots, N)$, and another certain symbol with length N sent by node 2 is $m_2[n]$. Which is, the symbol is the signal sent by node 1 is

$$x_1(t) = \sum_{n=1}^{N} m_1[n] \tag{8}$$

and node 2 send the message

$$x_2(t) = \sum_{n=1}^{N} m_2[n] \tag{9}$$

then the relay will receive

$$x_R(t, \varphi) = h_1 x_1(t) + h_2 x_2(t) + n(t) \tag{10}$$

where $h_1 = P_1 e^{\varphi_1}$ and $h_2 = P_2 e^{\varphi_2}$ with $P_1 = P_2$ when symmetrical channel is assumed. Then the message received at the relay can be rewritten as

$$x_R(t, \varphi) = P_0(x_1(t) + x_2(t)e^{\varphi}) + n(t) \tag{11}$$

where $P_0 = P_1 e^{\varphi_1}$ is a constant value.

Fig. 2. Sequence sent by node 1 and 2

After the relay receives $x_R(t, \varphi)$, it performs the mapping according to the mapping rule obtained in Sect. 2.2 to get

$$y_R(t) = x_1(t) \oplus x_2(t) \tag{12}$$

then sent back to node 1 and node 2. Note that the signal before transmission has already resolved the phase offset and noise at the relay, which requires the half-symbol asynchronization with belief propagation involved in Sect. 5. The mapped signal $y_R(t)$ is also composed of 16-QAM symbols, denoted as

$$y_R(t) = \sum_{n=1}^{N} m_R^y[n] \tag{13}$$

where

$$m_R^y = m_1 \oplus m_2 \tag{14}$$

3 Time Domain Phase Offset Problem for 16-QAM PNC

3.1 Constellation Pattern at the Relay in 16-QAM PNC

The following is based on the time-domain phase offset model of the 16-QAM PNC. Since this paper mainly studies the uplink, so we determine the BER by comparing the data received at the relay node with the data sent by the nodes 1 and 2. Therefore, it is of crucial importance whether the relay can receive the data correctly or not. In this case, we study the constellation pattern at the relay to illustrate the impact of phase offset.

Then the constellation at the 16-QAM PNC relay will be described. First, let us focus on the standard constellation at the relay, where only involves the phase offset and without noise. When performing PNC for 16-QAM, since there are 16 kinds of symbols sent by nodes 1 and 2 respectively, there are $16^2 = 256$ kinds of M_R involved at the relay. From Sect. 2.2 it can be seen that when $\varphi = 0$, one component of the signal at the relay makes a mapping of seven amplitudes to four amplitudes, so the overall signal at the relay should be mapped from 49 amplitudes to 16 amplitudes.

The constellation at the relay is shown in Fig. 3. In Fig. 3(a), the numbers in the upper right corner of each symbol represent the amount of the points overlapped in the same place, and the total sum is 256. The points in the figure are with various colors and shapes corresponding to the 16 types of 16-QAM symbol. It can be seen that when $\varphi = 0$, constellation points are of the same mapping coincide, and the demodulation can be the best at this time. When $\varphi = \pi/8$, the constellation pattern at the relay is shown in Fig. 3(b). It can be seen that some points belonging to different mappings are overlapped together, so it cannot be judged which symbol the constellation point at the overlapping position should be mapped to. Moreover, the middle part of the constellation point is too dense, which causes the small Euclidean distance between adjacent points, so it is prone to decode wrong.

When considering the noise ($snr = E_b/N_0 = 15\text{dB}$), the actual constellation at the relay is as shown in Fig. 4. It can be seen that when $\varphi = \pi/8$, the constellation points become even more dense and disorderly when $\varphi = 0$. The degree of point chaos in the middle part is very high, and the degree of agglomeration in the periphery points also decreases, thus a lot of code errors happen.

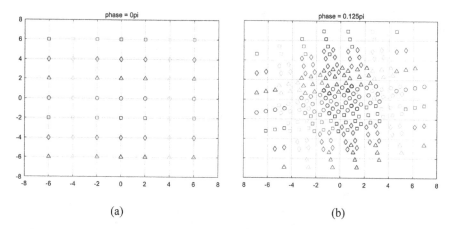

Fig. 3. Constellation pattern at the relay with no noise and (a) $\varphi = 0$ (b) $\varphi = \pi/8$

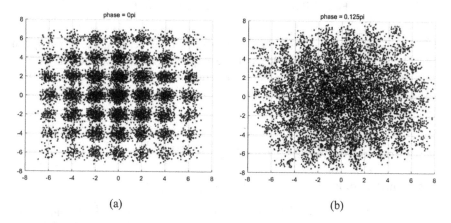

Fig. 4. Constellation pattern at the relay with $snr = 15\,\text{dB}$ and (a) $\varphi = 0$ (b) $\varphi = \pi/8$

3.2   16-QAM PNC Under Maximum Posterior Probability

Next, the Bit Error Rate (BER) curve will be drawn by MATLAB simulation in different SNR and different phase offsets. MAP decoding method is used in the simulation, that is, the Euclidean distance is determined between each received point and each point in the standard constellation map, and the point with the smallest Euclidean distance is selected as the mapping target. The SNR is sequentially incremented from 0 to 20 in steps of 1. In the same phase offset with different SNRs, each time node 1 and a node 2 send 1,000 data packets in sequence, and each data packet includes 10,000 16-QAM symbols.

Figure 5 shows the 16-QAM PNC BER performance curves for different phase shifts φ. It can be seen that when there is a small phase offset ($\varphi = \pi/8$), its performance is very poor already, and completely unable to meet the communication needs. When the phase shift increases to $\varphi = \pi/4$, the performance becomes even worse.

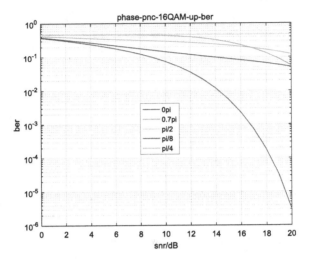

Fig. 5. BER curve at the relay in 16-QAM PNC

Thus, it can be seen that when there is phase offset, the performance of the system will deteriorate sharply and the BER will increase greatly. Even at large SNR, it cannot improve well and cannot meet the communication requirements. Although the MAP decoding can achieve the best reception, it cannot solve the phase shift decoding problem by itself.

4   Linear Channel Coding on 16-QAM PNC

4.1   Channel Coding Model Establishment

In general, channel coding has the ability of forward error correction. Using appropriate channel coding in Point to Point (P2P) channel can solve part of the transmission errors effectively and enhance the ability of noise or interference resisting, which can improve system reliability. The following will try to use linear channel coding in 16-QAM PNC and observe its performance changes.

As mentioned above, due to the assumption of symmetrical channels, the same channel coding scheme is used in all the links. Let the original bit information at nodes 1 and 2 be channel-coded before being sent out. As described in Sect. 2.3, one of the original symbol sent by node 1 or node 2 is $m_1[i]$ $(n = 1, 2, \ldots, i, \ldots, N)$ or $m_2[i]$, respectively. Let a linear channel coding method (function) be C. Symbol $m_1[i]$ and $m_2[i]$ are coded to $C(m_1[i])$ and $C(m_2[i])$, then sent out at the same time. Then the relay will receive the signal

$$x_R(t) = x_1(t) + x_2(t) = C(\sum_n m_1[n]) + C(\sum_n m_2[n]) \tag{15}$$

Now there are two ways to deal with this $x_R(t)$.

(a) Due to

$$C(m_1[i]) + C(m_2[i]) = C(m_1[i] + m_2[i]) \tag{16}$$

the relay will receive the signal, channel decode it to $m_1[i] + m_2[i]$, and then perform mapping mentioned above to get

$$m_R^y[i] = m_1[i] \oplus m_2[i] \tag{17}$$

Then channel code it to $C(m_R^y[i])$ and send it to node 1 and 2.

(b) Due to linear channel coding method being used here, the relay can perform PNC on $C(m_1[i]) + C(m_2[i])$ to get $C(m_1[i]) \oplus C(m_2[i])$. And obviously,

$$C(m_1[i]) \oplus C(m_2[i]) = C(m_1[i] \oplus m_2[i]) \tag{18}$$

So $m_1[i] \oplus m_2[i]$ is get after channel decoding.

(c) Multi-User Detection (MUD) technology can be used to achieve the detection of the original signal (such as $x_1(t)$ or $x_2(t)$), and then the corresponding symbol finish XOR operation to get $\sum_n m_1[n] \oplus m_2[n]$. This will increase the system burden, because relays do not need to know what $x_1(t)$ or $x_2(t)$ is. The relay is only concerned about getting $m_1[n] \oplus m_2[n]$ correctly.

It has been pointed out in the paper [1] that the method (a) is better because channel decoding can solve some of the channel mapping errors. In the method (b) which is directly perform PNC mapping on $C(m_1[i]) + C(m_2[i])$, the error brought by the channel will be passed on and cannot be resolved; that is, if $C(m_1[i]) + C(m_2[i])$ is wrong, then the mapping will be wrong. This kind of mistake cannot be corrected by channel decoding. Therefore, this section uses method (a) for 16-QAM PNC to achieve channel coding.

In this way, after the steps in (a), the terminal node (with node 1 as an example) receives the signal $C(m_1[i] \oplus m_2[i])$, then calculates

$$m_1[i] \oplus (C^{-1}(C(m_1[i] \oplus m_2[i])) = m_2[i] \tag{19}$$

to get the message sent by node 2.

4.2 Polar Coding on 16-QAM PNC

Different channel coding methods also have different error correction capabilities. Since this paper does not regard channel coding as a variable, this paper fixes a representative channel coding method to do related research. Polar code was first proposed by Arikan in 2008 [11]. Its practicability is very high and meets the requirements of this article. In the following, a polar code with the code rate $R = 0.5$ and the code length of 8 is described.

"Polar code" is called because it deriving the channel polar to construct a codeword that can achieve symmetric channel capacity. Note that for an N bit channel W_N, the symmetric channel capacity of a binary discrete memoryless channel is given by

$$I(W_N) = \sum_{y \in Y} \sum_{x \in X} \frac{1}{2} W(y|x) \log \left(\frac{2W(y|x)}{W(y|0) + W(y|1)} \right) \tag{20}$$

where x is the channel input, y is the channel output, the value of $I(W_N)$ is $[0, 1]$. Bhattacharyya parameter in the W_N is given by

$$Z(W_N) = \sum_{y \in Y} \sqrt{W(y|0)W(y|1)} \tag{21}$$

and its value is between 0 and 1. $Z(W_N)$ measures the channel reliability.

When various circumstances' probability of the input W_N is equal, $I(W_N)$ reach the maximum value. It can be intuitively obtained if and only if there is $Z(W_N) \approx 0$ then $I(W_N) \approx 1$, and only when $Z(W_N) \approx 1$ then $I(W_N) \approx 0$.

An important process of polar coding is to choose a reliable channel to convey useful information. In general, the generator matrix of the polar codes with of the block size $N = 2^n$ is represented by

$$\mathbf{G}_N = \mathbf{F}^{\otimes n} \tag{22}$$

and $\mathbf{F} = \begin{pmatrix} 1 & 0 \\ 1 & 1 \end{pmatrix}$, where $\mathbf{F}^{\otimes n}$ represents the Kronecker inner product. The channel codeword with bit length N is given by

$$\mathbf{w} = \mathbf{u}\mathbf{G}_N \tag{23}$$

where $\mathbf{u}$ is the input of the channel coder. In this paper the polar code's length is $N = 8 = 2^3$ and code rate is $R = 0.5$, so the input $\mathbf{u}$ is an 1×8 vector, which contains a 4 bit higher $I(W)$ and lower "pureness" channel as the channel to send information, in which the arrangement of information bits; and 4 bit higher $Z(W)$ and lower "clutter" channel as the noise channel, in which the fixed bit arranged.

In order to select a 4-bit reliable channel, it is necessary to calculate the value of the channel capacity $I(W)$. According to the correlation calculation method of the polar coding [11], it can be obtained that

$$I(W_{N=8}) = \{0.0039, 0.1211, 0.1914, 0.6836, 0.3164, 0.8086, 0.8789, 0.9961\}$$

It can be seen that the corresponding values of the $I(W)$ at the fourth, sixth, seventh and eighth bits are larger, and their corresponding positions should be placed in the information bits. Then, according to the specific process of the general polar code coding, the channel arrangement should be done according to the following method.

In the transmitter, an 8-bit encoding input polar code should be arranged the fourth, sixth, seventh, and eighth bits used as the information bits. In the simulation of this paper, the fixed position is all placing 0; it can be proved that in the symmetric channel, the value of the fixed bits has no effect on decoding. At the receiving end, the codeword is decoded through Successive Cancellation (SC) decoding to get specific information whose length is 8, of which the seventh, sixth, fourth, and eighth bits correspond to first, second, third and fourth bits of the effective decoding information, respectively. For example, for a 4-bit codeword $\{1,1,0,1\}$, the coded codeword is $\{1,1,0,0,0,0,1,1\}$. In this way, the coding and decoding process of the entire polar code is completed.

Using polar coding in the case of different phase migration φ, the system performance of 16-QAM PNC is shown in Fig. 6. In this simulation, the SNR is from 0 to 20 in 1 step length. Under each SNR condition with different phase offset, nodes 1 and node 2 send 500 packets in turn, and each packet contains 2500 16-QAM symbols. From Fig. 6, it can be seen that the polar code channel coding does not improve the system performance of 16-QAM in PNC phase shift. As a result of encoding a single species here, it cannot be concluded like "channel encoding of 16-QAM PNC is useless", but in view of the polar code is a powerful error correcting capability of encoding, channel encoding itself is lacking the possibility to the problem of the phase shift in 16-QAM PNC.

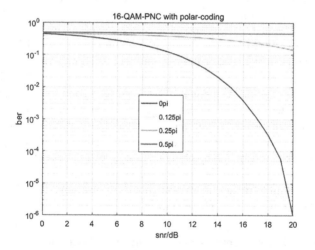

Fig. 6. BER of 16-QAM PNC using polar coding

5 Half Symbol Asynchronization with Belief Propagation

5.1 The Proposal of the Half-Symbol Asynchronous Belief Propagation

Section 3 summarizes the general case of 16-QAM PNC performance. It can be seen in Fig. 3(b) that, in the case of 256 points scattered unevenly within the constellation pattern, where the inner constellation points are relatively dense and the Euclidean distance between them is relatively small. If two symbols' sum $m_1[a] + m_2[b]$ is mapped to the inside, it is more difficult to make the right judgment. In general, there is a barrel effect in the system's BER [1], that is, the constellation point at which the system's SNR receives the worst-case error, as if the ability to hold water in the barrel was constrained by the shortest plank. Therefore, the decision of the inner mapping is difficult and error prone, resulting in a high system error and cannot easily be improved.

However, the outer constellation points are relatively scattered, and it is relatively easy to make a correct judgment to $m_1[a] + m_2[b]$ who is mapped to the outside. Notice that for every constellation symbol in Fig. 3(b), once the relay knows that it should map to a specific point, its corresponding $m_1[a]$ and $m_2[b]$ sent by node 1 or node 2 will be known. Then, as the point mapping to the outside of the constellation has a larger probability to make a correct decision, there is a larger probability of getting the two information at the same time. If this information can be utilized to help demodulate the points that are mapped to the inside side, it will be beneficial to improve the overall BER [12].

Belief Propagation (BP) is an algorithm for the implementation of the probabilistic computing problem containing the message transfer process. An undirected graph model is constructed as Markov random field, where each point represents a random variable, and the belief propagation algorithm is a probabilistic inference method based on it. If some information of the points in the random field is known, it is necessary to get some other information using this information. For every node in the field, the probability state of a node is transmitted to another adjacent node by message propagation, and the probability state at another node is updated. Information is transmitted repeatedly and updated at all points through repeated iterations. After several iterations, the probability distribution at each point will tend to a steady state. That is to say, the random field has reached the convergence state, and each point has the best confidence. In general, it is to find the edge probability distribution of each point in the random field. In this way, the overall performance of the system can be promoted.

To realize the previous idea, this paper proposes a design of two nodes in the uplink sending the information with half symbol duration asynchronous between each other in purpose. Then the relay receives the asynchronous signal and performs the belief propagation algorithm to obtain the correct codeword mapping. The simulation shows that it can solve the problem of phase shift in 16-QAM PNC system.

5.2   Half-Symbol Asynchronous System and the BP Algorithm

As described in Sect. 2.3, the signal sent by node 1 is $x_1(t) = \sum_1^N m_1[n]$, and node 2 is $x_2(t) = \sum_1^N m_2[n]$, where m is a 16-QAM symbol. Assuming that the sequence of symbols transmitted by node 1 arrives at relay R a half-symbol earlier than node 2. For the convenience of discussion, the end of the $x_1(t)$ is complemented of a half-symbol $m_1[N+1]$ which does not actually exist, and in the front of the $x_2(t)$ a half-symbol

$m_2[0]$ is complemented which actually does not exist either. Now the length of $x_1(t)$ and $x_2(t)$ is equal. Then $x_R(t)$ at this time will be

$$m_R[a+b] = m_1[a] + m_2[b] \tag{24}$$

where $a \in \{1, 2, 3, \ldots, N + 1\}$, $b \in \{0, 1, 2, 3, \ldots, N\}$ with $a - b \leq 1$; then

$$x_R(t) = \sum_{n=1}^{2N+1} m_R[n] \tag{25}$$

Figure 7 shows the details of a half-symbol asynchrony system. In Fig. 7, due to asynchrony, the sample value has been increased to $2N + 1$, which the relay can obtain more information comparing the synchronization case. For $a = b (m_R[4] = m_1[2] + m_2[2]$ as example), it corresponds to the sampling information at the synchronous situation, which is also the key information needed to derive $m_R^y = m_1 \oplus m_2$; for $a - b = 1$ $(m_R[7] = m_1[4] + m_2[3]$ as example), it corresponds to the sampling information at the asynchronous situation, which is the "more" information get here. Whether $m_R^y = m_1 \oplus m_2$ is correct or not can be judged through the posterior probability

$$P(m_R^y = m_1 \oplus m_2 | m_R) \tag{26}$$

This probability is the confidence level to be calculated. In this way, we construct the Markov random field needed by this paper, as shown in Fig. 8. The right-hand side of the $x_R(t)$ shows the $2N + 1$ values sampled by the relay after receiving, and these values are used to calculate the confidence probability. η means check node which provides external information decoding process, whose verification rule is

$$\eta(a, b) = \begin{cases} 1, m_R[a] = m_R[b] \\ 0, m_R[a] \neq m_R[b] \end{cases} \tag{27}$$

Where $a \in \{1, 2, 3, \ldots, N + 1\}$, $b \in \{0, 1, 2, 3, \ldots, N\}$, $a - b \leq 1$.

The general process of information update in BP algorithm is to update the probability information iteratively to calculate the best confidence of each sample point (i.e., the maximum value of Eq. (26)), and make corresponding mapping according to the confidence judgment result to get the corresponding $m_R^y(t)$ on the top.

BP algorithm is a relatively mature algorithm, its information iterative updating process is basically fixed [9].

(a) information initialization, including the likelihood function of all hidden nodes, and potential energy and message value of each neighbor node;
(b) update of the check node;
(c) update of the value of the bit node;
(d) making judgments, verification. If correct, stop decoding; otherwise return to step (a).

The specific details of the BP algorithm are not the focus of this article, so we do not do in-depth discussion here.

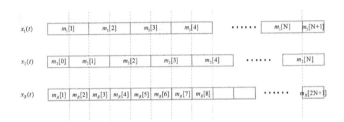

Fig. 7. Half-symbol asynchronous system

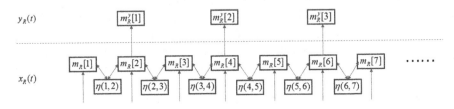

Fig. 8. Confidence propagation diagram

5.3 The Capability of Half-Symbol Asynchronous BP Algorithm

The half-symbol asynchronous BP algorithm proposed in this paper can reflect its good performance in the following two aspects.

(a) Half-symbol asynchronous. In the synchronization case of Fig. 2, the symbols $x_1(t)$ and $x_2(t)$ are corresponded one by one. When the relay receives $x_R(t)$, it samples and obtain N samples. In Fig. 7, the sample values for each are doubled because of asynchrony, and the original information obtained at the relay is doubled (2N + 1 samples in total), allowing for more accurate decoding and mapping decisions.

(b) Information transfer implied in BP algorithm (Fig. 9). As mentioned in Sect. 5.2, the BP decoding algorithm is the finding the maximum of the probability of Eq. (26); in all sample values of $x_1(t) + x_2(t)$, values like

$$m_R[2i] = m_1[i] + m_2[i] \tag{28}$$

is the key to launching the required mapping. If it is mapped to the inner position in Fig. 3(b), it will be difficult to determine the constellation attribution correctly, because

$$P(m_R^y[i] = m_1[i] \oplus m_2[i] | m_R[2i]) \tag{29}$$

its maximum is small and with low confidence. However, the sample values adjacent to it, which are

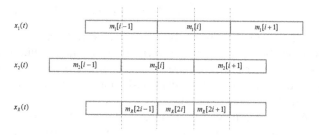

Fig. 9. Information transfer in BP algorithm

$$m_R[2i+1] = m_1[i+1] + m_2[i] \tag{30}$$

or

$$m_R[2i-1] = m_1[i] + m_2[i-1] \tag{31}$$

may be mapped to the outer position of the constellation, and the probability of correct demodulation will be higher. Assuming Eq. (31) is correctly demodulated, then $m_1[i]$ and $m_2[i-1]$ can be known, and they can help to demodulate $m_R[2i]$, which causes the sample value being successfully solved to the correct information because of the high confidence of $m_1[i]$ and $m_2[i-1]$. In this way, BP algorithm can enhance the overall decoding system.

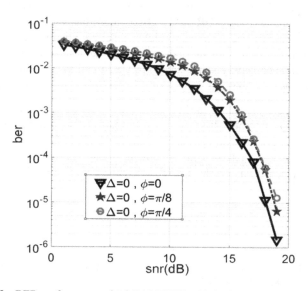

Fig. 10. BER performance of 16-QAM PNC with half-symbol BP algorithm

5.4   Simulation on 16-QAM PNC with Half-Symbol BP Algorithm

By generating half-symbol asynchrony BP algorithm, the performance of the 16-QAM PNC system at different phase offsets is shown in Fig. 10. In the simulation here, the SNR progressively advances from 0 to 20 in steps of 1. Under the same condition and different SNR, node 1 and node 2 simultaneously transmit 100 data packets in turn, each containing 1,000 16-QAM symbols.

In Fig. 10, Δ is a half-symbol asynchronous parameter. $\Delta = 0$ means half-aligned, and $0 \leq \Delta < 0.5$. ϕ means the phase offset φ. As can be seen, the BER at $\varphi = \pi/8$ is 6 dB higher than $\varphi = 0$ when SNR is 18 dB, and the performance between $\varphi = \pi/4$ and $\varphi = \pi/8$ is not much difference. Compared with Fig. 5, the proposed decoding method solves the problem of phase offset, and when the SNR is large ($snr = 18$ dB), the BER performance is improved by 34 dBs ($\varphi = \pi/4$) and 30.7 dB ($\varphi = \pi/8$), which means the result can meet the normal communication needs. So, it can be seen that this algorithm has very good performance.

6   Conclusion

In this paper, we study the phase asynchrony issue in 16-QAM modulated PNC system, including the establishment and mapping of 16-QAM PNC, the general performance of system, system performance combined with channel coding. In the situation that the traditional idea cannot solve this problem, this paper proposes a half-symbol asynchronous BP algorithm to solve the problem caused by the phase offset by constructing asynchronously half symbols using the belief propagation algorithm.

The essence of PNC is to achieve the relay at the source of two data packets to XOR operation. In this paper, the mapping rule of 16-QAM relay has been successfully constructed, and the bitwise XOR at 16-QAM relay has been implemented. This method is general and can be extended to higher order modulations (such as 64-QAM, 1024-QAM, etc.). Moreover, for the half-symbol asynchronous BP algorithm in this paper, it has been successfully achieved on 16-QAM and the performance is greatly improved. It can be seen that this algorithm still has space for further expansion.

Future Work. There are some further discussions can be done in the future. For example, the way to combine half-symbol asynchronous BP algorithm with the existing means of communication technology (such as channel coding, OFDM, etc.) is still need for further study. For the research in this article, it can be continued to simulate the actual situation (such as using the USRP to do the actual simulation) to get more accurate results.

References

1. Liew, S.C., Lu, L., Zhang, S.: A primer on physical-layer network coding. Synth. Lect. Commun. Netw. **8**(1), 1–218 (2015)
2. Zhao, M.F., Zhou, Y.J., Yuan, Q., Yang, Y.X.: Research survey on physical layer network coding. J. Comput. Appl. **31**(8), 2015–2128 (2011)

3. Zhang, S., Liew, S.C., Lam, P.P.: Physical-layer network coding. In: International Conference on Mobile Computing and Networking, MOBICOM 2006, pp. 358–365 (2006)

4. Chen, L.: The research of physical layer network coding strategy based on various modulation techniques. Beijing Institute of Technology (Doctoral dissertation) (2015)

5. Liew, S.C., Zhang, S., Lu, L.: Physical-layer network coding: tutorial, survey, and beyond. Phys. Commun. **6**, 4–42 (2011)

6. Pan, H., Lu, L., Liew, S.C.: Network-coded multiple access with high-order modulations. In: IEEE Global Communications Conference, pp. 1–7 (2016)

7. Du, J., Yang, L., Yuan, J., Zhou, L., He, X.: Bit mapping design for LDPC coded BICM schemes with multi-edge type exit chart. IEEE Commun. Lett. **21**(4), 722–725 (2017)

8. Lu, L., Wang, T., Liew, S.C., Zhang, S.: Implementation of physical-layer network coding. Elsevier Phys. Commun. **6**(1), 74–87 (2013)

9. Goldsmith, A.: Wireless Communications. Cambridge University Press, Cambridge (2005)

10. Leon-Garcia, A., Widjaja, I.: Communication Networks. Tsinghua University Press, Beijing (2000)

11. Arikan, E.: Channel polarization: a method for constructing capacity-achieving codes for symmetric binary-input memoryless channels. IEEE Trans. Inform. Theory **55**(7), 3051–3073 (2009)

12. Lu, L., Liew, S.C.: Asynchronous physical-layer network coding. IEEE Trans. Wirel. Commun. **11**(2), 819–831 (2012)

Non-orthogonal Multiple Access for Similar Channel Conditions

Asim Anwar, Boon-Chong Seet, and Xue Jun Li$^{(\boxtimes)}$

Department of EEE, Auckland University of Technology, Auckland, New Zealand
{aanwar,bseet,xuejun.li}@aut.ac.nz

Abstract. Non-orthogonal multiple access (NOMA) is considered as promising multiple access (MA) scheme for upcoming fifth generation (5G) systems. The performance of NOMA is highly dependent upon having significant channel gain difference among users. In this paper, we focus on the situation of similar channel conditions and propose a channel gain stretching (CGS) strategy to apply NOMA more effectively under these conditions. In order to evaluate the performance, we derive a closed-form expression of the outage probability. Numerical results are also presented to validate the accuracy of the derived results and also to compare the performance of NOMA with and without CGS, and orthogonal MA (OMA).

Keywords: Non-orthogonal multiple access
Channel gain stretching · Power allocation

1 Introduction

In recent years, there has been a tremendous growth in the number of cellular subscribers coupled with the massive increase in portable devices to experience diverse services spanning from simple voice to high data rate real time multimedia applications. Moreover, this explosive growth in mobile devices requires a vigorous demand of seamless and ubiquitous connectivity. In order to meet the anticipated demands of future fifth generation (5G) communication systems, these trends pose a major challenge to network operators due to the scarcity of current spectrum resources [1].

Many potential solutions are proposed to realize the concept of 5G among which millimeter waves, massive multiple input multiple output (MIMO), full-duplex, heterogeneous deployments and software defined networks have received notable attention from both academia and industry [2]. Nevertheless, multiple access (MA) scheme always play a critical role to enhance spectrum efficiency in a cost-effective manner.

Non-orthogonal multiple access (NOMA) has been considered as a latest member of MA family and is proposed as a promising MA technology for 5G systems. The key idea of NOMA is that it superimposes multiple users into single resource (time/frequency/code) at the transmitter side by allocating different

© ICST Institute for Computer Sciences, Social Informatics and Telecommunications Engineering 2018
P. H. J. Chong et al. (Eds.): SmartGIFT 2018, LNICST 245, pp. 192–199, 2018.
https://doi.org/10.1007/978-3-319-94965-9_19

power levels to each user and applies successive interference cancellation (SIC) at the user's receiver to mitigate intra-user interference. Some of the key benefits of deploying NOMA include improved spectral efficiency, enhanced throughput and better fairness among users [3].

1.1 Related Work, Motivation and Contributions

The initial investigations on NOMA were conducted in [4] via system level simulations. The authors reported superior throughput and performance of NOMA over conventional orthogonal MA (OMA) scheme. The outage performance of NOMA with randomly deployed users is analytically derived and then evaluated in [5]. The application of MIMO systems to NOMA is explored in [6]. The authors presented novel design of precoder which is then utilized to suppress the inter-beam interference. The impact of user pairing on the performance of NOMA system is investigated in [7]. The authors discussed and evaluated the performance of two possible implementations of NOMA systems, namely fixed power allocation NOMA and cognitive-radio-inspired NOMA (CR-NOMA). A NOMA-based device-to-device (D2D) communication is proposed in [8] with underlay cellular network. The concept of group D2D communications is introduced in which D2D transmitter is communicating with multiple D2D receivers via NOMA protocol. In order to manage the interference from underlying uplink cellular communication, an optimal resource allocation strategy was proposed.

More recently, cooperative NOMA is proposed in [9] where strong user is equipped with full-duplex functionality. The authors proposed a scheme to improve the outage performance of a weak user using cooperative and direct transmissions by invoking D2D communications between strong and weak NOMA user pair. A large-scale D2D network is considered in [10], where the authors proposed a cooperative hybrid automatic repeat request assisted NOMA scheme to improve the outage and throughput performance of the D2D users.

In all the aforementioned works, the underlying assumption is to maintain a significant channel gain difference among NOMA users. However, this assumption may not always hold and under those scenarios, it may result in improper rate and power allocation that could result in complete outage [5]. This motivates us to propose a method that artificially generates a channel gain difference among different NOMA users for proper power allocation under situations of similar channel conditions. To this end, the main contributions of this work are summarized below:

- We find minimum optimal power allocation coefficients that would guarantee to meet the targeted rate of each user.
- We propose a channel gain stretching (CGS) scheme to apply NOMA effectively under comparable channel conditions.
- In order to evaluate performance, exact expressions for outage probability are derived.
- Numerical results are shown to validate the accuracy of the analysis, as well as compare outage performance of the NOMA under proposed CGS scheme to NOMA without CGS and OMA.

2   System Model

Consider a single-input single-output system with single source (S) located at the center of a disc D with radius R_D. We focus on a downlink scenario where S is communicating with M users via NOMA protocol. The users are randomly uniformly distributed inside disc D. The channel gain between user m and source S is given as, $h_m = g_m d_m^{-\alpha}$, where g_m is the power fading coefficient that follows exponential distribution with unit mean, d_m is the distance between user m and source S and α is the path loss exponent.

In this work, we consider a scenario of similar channel conditions. These situations may arise in practical scenarios that could include (1) Indoor deployments with source implementing NOMA and where channel conditions for users are expected to be very similar, (2) NOMA based group D2D communication forming a small local cell where the users exit in proximity of each other and are clustered around group head/transmitter [8] and (3) User-centric deployments of small-cell base stations where the users are clustered around small-cell base stations. Under these kind of situations, all M NOMA users have similar channel conditions i.e. $h_i \approx h_j, i \neq j, 1 \leq i, j \leq M$. It should be noted here that the channel gains of all users are not exactly the same and hence user ordering is still possible. Without loss of generality, the users are ordered as $h_1 \leq ... \leq h_M$. Consequently, the power allocation coefficients, denoted as, $a_m, 1 \leq m \leq M$, and are sorted as, $a_1 \geq ... \geq a_M$. The calculation of power allocation coefficients is discussed in the next sub-section.

2.1   Minimum Required Power Allocation Coefficients

Let us denote R_m and $\bar{R}_m$ by achievable and targeted rates of the user m respectively. Then, $\bar{R}_m$ of user m is met if:

$$R_m \geq \bar{R}_m. \tag{1}$$

Equation (1) can be further simplified as:

$$\log_2 \left(1 + \frac{P h_m a_m}{P h_m \sum_{i=m+1}^{M} a_i + \sigma^2} \right) \geq \bar{R}_m$$

$$a_m \geq \tau_m \left(\sum_{i=m+1}^{M} a_i + \frac{1}{\Upsilon h_m} \right), \tag{2}$$

where $\tau_m = 2^{\bar{R}_m} - 1$, $\Upsilon = \frac{P}{\sigma^2}$ is the transmit signal-to-noise ratio (SNR), P is the maximum transmit power at the base station and σ^2 is the variance of additive noise. In order to proceed forward, we formulate the following optimization problem to obtain the optimal power allocation coefficients.

$$\min \sum_{m=1}^{M} a_m \tag{3}$$

$$s.t.(2) \qquad\qquad\qquad (4)$$

To this end, the following lemma states the optimal power allocation coefficients that are sufficient to meet the users' targeted rates.

Lemma 1. *The optimal power allocation coefficients are obtained by solving the problem (3) and are given as:*

$$a_m = \tau_m \left(\sum_{i=m+1}^{M} a_i + \frac{1}{\varUpsilon h_m} \right) \qquad (5)$$

Proof. By inspecting problem (3), it can be observed that (3) is convex. Hence, a necessary and sufficient condition to obtain its optimal solution follows by the application of Karush-Kuhn-Tucker (KKT) conditions. The detailed proof follows a standard application of KKT conditions and hence is skipped. Curious reader is referred to see a Theorem 1 [11] for the detailed proof. $\qquad\square$

3   Proposed Channel Gain Stretching Method and Outage Analysis

In this section, we first propose a CGS method that artificially generates channel gain difference among different NOMA users. Then, under the proposed CGS scheme, we analyse the outage probability of the considered system.

3.1   Channel Gain Stretching Method

Under situations of similar channel conditions, we propose a following transformation to artificially generate channel gain difference among NOMA users:

$$\bar{h}_m = k_{1,m} \left(h_m \right)^{k_{2,m}}, \qquad\qquad (6)$$

where $\bar{h}_m$ is the transformed channel gain of user m and $k_{1,m} > 0, k_{2,m} > 0$ are positive constants for user m and are selected in such a way to achieve a significant difference among channel gains of users.

Example: Consider a case of two users with $(h_1, h_2) = (0.87, 0.9)$. Now applying (6) with $k_{1,1} = k_{2,1} = 0.5, k_{1,2} = 3, k_{2,2} = 3.5$ results in stretched coefficients as $\left(\bar{h}_1, \bar{h}_2 \right) = (0.46, 2)$. The power allocation coefficients are then computed using (5) for a given SNR and targeted rate.

3.2   Outage Analysis

The outage occurs at the user m receiver whenever it fails to decode the message signal of any higher order user $j, 1 \leq j \leq m$. Then, the outage probability of user m in decoding user j can be expressed as:

$$P_{m \to j} = \Pr\left(\frac{\bar{h}_j a_j \Upsilon}{\bar{h}_j \Upsilon \sum_{i=m+1}^{M} a_i + 1} < \tau_j\right)$$

$$= \Pr\left[\bar{h}_j \Upsilon \left(a_j - \tau_j \sum_{i=m+1}^{M} a_i\right) < \tau_j\right]$$

$$= \Pr\left(\bar{h}_j < \frac{\tau_j}{\Upsilon \left(a_j - \tau_j \sum_{i=m+1}^{M} a_i\right)}\right)$$

$$= \Pr\left[h_j < \left(\frac{\varphi_j}{k_{1,j}}\right)^{k_{2,j}}\right]$$

$$= F_{h_j}(\theta_j), \tag{7}$$

where $\varphi_j = \frac{\tau_j}{\Upsilon\left(a_j - \tau_j \sum_{i=m+1}^{M} a_i\right)}$, $\theta_j = \left(\frac{\varphi_j}{k_{1,j}}\right)^{k_{2,j}}$ and F_{h_j} is the cumulative distribution function (CDF) of h_j. Now let us define $\theta_m^{\max} = \max\{\theta_1, ..., \theta_m\}$. The outage probability at user m is then given as:

$$P_m = F_{h_m}(\theta_m^{\max}). \tag{8}$$

In order to obtain outage probability P_m of user m, we require CDF of h_m which is obtained by analyzing order statistics [12] and is given as:

$$P_m = \mu_m \sum_{l=0}^{M-m} \binom{M-m}{l} (-1)^l \int_0^{\theta_m^{\max}} \left(F_{\hat{h}}(x)\right)^{m+l-1} f_{\hat{h}}(x)dx, \tag{9}$$

where $F_{\hat{h}}$ and $f_{\hat{h}}$ are the CDF and probability density function (PDF) of the unordered channel gain $\hat{h}$ respectively. The CDF $F_{\hat{h}}$ of the unordered channel gain is given as [13]:

$$F_{\hat{h}}(x) = \frac{2}{R_D^2} \int_0^{R_D} \left(1 - e^{-(1+z^\alpha)x}\right) z dz$$

$$\stackrel{(a)}{=} \frac{\delta}{R_D^2} \int_0^{R_D^\alpha} \left(1 - e^{-(1+y)x}\right) y^{\delta-1} dy$$

$$\stackrel{(b)}{=} 1 - \delta e^{-x} \mathrm{B}(1, \delta) \Phi(\delta, 1+\delta; -xR_D^\alpha), \tag{10}$$

where (a) and (b) are obtained by a change of variable from $z^\alpha \to y$ and applying Eq. 3.383 of [14] respectively, $\delta = \frac{2}{\alpha}$, $\mathrm{B}(\cdot, \cdot)$ is the beta function and $\Phi(\cdot, \cdot; \cdot)$ is the confluent hypergeometric function. Now, take the derivative of (10) to obtain $f_{\hat{h}}$ and substitute $F_{\hat{h}}$ and $f_{\hat{h}}$ in (9), P_m can be expressed as:

$$P_m = \mu_m \sum_{l=0}^{M-m} \binom{M-m}{l} (-1)^l \int_0^{\theta_m^{\max}} \delta B(1,\delta) e^{-x} [\Phi(\delta, 1+\delta; -xR_D^\alpha)$$

$$+ \rho\Phi(1+\delta, 2+\delta; -xR_D^\alpha)]$$
$$\times [1 - \delta e^{-x} B(1,\delta) \Phi(\delta, 1+\delta; -xR_D^\alpha)]^{m+l-1} \, dx. \quad (11)$$

The analytical solution of (11) is difficult to obtain and hence we apply Gaussian-Chebyshev quadrature to approximate the outage probability of user m as follows:

$$P_m = \mu_m \sum_{l=0}^{M-m} \binom{M-m}{l} (-1)^l \left\{ \sum_{n=1}^{N} \Psi_n [\Phi(\delta, 1+\delta; -b_n) \right.$$

$$+ \rho\Phi(1+\delta, 2+\delta; -b_n)]$$
$$\left. \times [1 - \delta e^{-\theta_m^{\max} s_n} B(1,\delta) \Phi(\delta, 1+\delta; -b_n)]^{m+l-1} \right\}, \quad (12)$$

where $b_n = \theta_m^{\max} s_n R_D^\alpha$, $s_n = \frac{1}{2}(1 + \vartheta_n)$, $\vartheta_n = \cos(\frac{2n-1}{2N}\pi)$, $\Psi_n = \delta \omega_n \sqrt{1 - \vartheta_n^2} \, B(1,\delta) \, \theta_m^{\max} e^{-\theta_m^{\max} s_n}$, $\omega_n = \frac{\pi}{N}$ and N is the complexity-accuracy trade-off parameter.

4   Numerical Results

This section presents the numerical simulations to validate the accuracy of derived outage results as well as to compare the performance of NOMA system under proposed CGS scheme with no CGS applied and OMA by considering similar channel conditions for all users. In all simulations, we consider $M = 2$, $\bar{R}_1 = \bar{R}_2 = 1$ bits per channel use, $R_D = 20$ m, $\Upsilon = [10 - 50]$ dB and $N = 5$. Further, as a representative case, the parameters $\{h_m, \bar{h}_m, k_{1,m}, k_{2,m}\}_{m=1}^{M}$ are taken from Example (Sect. 3.1).

The impact of varying R_D on the outage performance of the users is presented in Fig. 1. Following observations are drawn from the results. First, increasing the radius R_D increases the outage probability of the users due to the higher path loss. Second, user $m = 1$ has lower outage probability than user $m = 2$ because under similar channel conditions scenario, the application of CGS results in $\bar{h}_1 < \bar{h}_2$ (see Example in Sect. 3.1). As a consequence, $a_1 > a_2$ for all SNR values which results in better performance of user $m = 1$. Moreover, Monte-Carlo simulations are also performed to validate the accuracy of derived results in (12). It can be observed that the analytical and simulation results are in good agreement.

The outage performance among NOMA system with CGS, without CGS and OMA is presented in Fig. 2. It can be observed that NOMA under proposed CGS scheme outperforms NOMA without CGS and OMA. Further, it can be noted that the performance of NOMA without CGS is badly impacted. These results can be explained as follows: The scenario of similar channel conditions result in

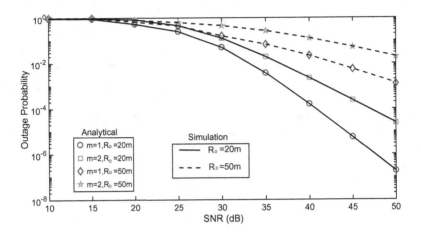

Fig. 1. Impact of varying R_D on outage performance.

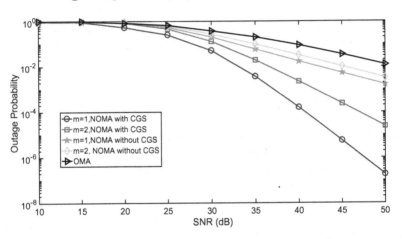

Fig. 2. Outage performance comparison among NOMA with and without CGS and OMA.

very comparable power allocation coefficients, which then increase the signal-to-interference-plus-noise ratio (SINR) threshold required for SIC decoding. By applying proposed CGS using (6) produces significant difference in channel gains resulting in significantly different power allocation coefficients and hence reducing the SINR threshold for SIC decoding. Further, both $m = 1, 2$ users have similar channel conditions, therefore, application of OMA results in same performance for both users, and hence we presented only one result for OMA scheme.

5 Conclusion

In this work, we consider a scenario of similar channel conditions for NOMA. In order to apply NOMA more effectively under these situations, we propose CGS method to artificially generate a channel gain difference among users. Closed-form expression for outage probability is derived to characterize the performance. It can be observed from the results that the NOMA under proposed CGS method outperforms NOMA without CGS and OMA under situations of comparable channel conditions. As future extension of this work, we plan to extend the proposed scheme for MIMO systems.

References

1. Akyildiz, I.F., Nie, S., Lin, S.C., Chandrasekaran, M.: 5G roadmap: 10 key enabling technologies. Comput. Netw. **106**(Suppl. C), 17–48 (2016)
2. Wong, V.W.S., Schober, R., Ng, D.W.K., Wang, L.C.: Key Technologies for 5G Wireless Systems. Cambridge University Press, Cambridge (2017)
3. Ding, Z., Liu, Y., Choi, J., Sun, Q., Elkashlan, M., Chih-Lin, I., Poor, H.V.: Application of non-orthogonal multiple access in LTE and 5G networks. IEEE Commun. Mag. **55**(2), 185–191 (2017)
4. Saito, Y., Benjebbour, A., Kishiyama, Y., Nakamura, T.: System-level performance evaluation of downlink non-orthogonal multiple access (NOMA). In: Proceedings of IEEE 24th Annual International Symposium on Personal, Indoor, and Mobile Radio Communications (PIMRC), London, UK, September 2013
5. Ding, Z., Yang, Z., Fan, P., Poor, H.V.: On the performance of non-orthogonal multiple access in 5G systems with randomly deployed users. IEEE Sig. Process. Lett. **21**(12), 1501–1505 (2014)
6. Ding, Z., Adachi, F., Poor, H.V.: The application of mimo to non-orthogonal multiple access. IEEE Trans. Wirel. Commun. **15**(1), 537–552 (2016)
7. Ding, Z., Fan, P., Poor, H.V.: Impact of user pairing on 5G nonorthogonal multiple-access downlink transmissions. IEEE Trans. Veh. Technol. **65**(8), 6010–6023 (2016)
8. Zhao, J., Liu, Y., Chai, K.K., Chen, Y., Elkashlan, M., Alonso-Zarate, J.: Noma-based D2D communications: towards 5G. In: 2016 IEEE Global Communications Conference (GLOBECOM), Washington, D.C., USA, December 2016
9. Zhang, Z., Ma, Z., Xiao, M., Ding, Z., Fan, P.: Full-duplex device-to-device-aided cooperative nonorthogonal multiple access. IEEE Trans. Veh. Technol. **66**(5), 4467–4471 (2017)
10. Shi, Z., Ma, S., ElSawy, H., Yang, G., Alouini, M.: Cooperative HARQ assisted NOMA scheme in large-scale D2D networks. CoRR abs/1707.03945 (2017)
11. Zhang, Y., Wang, H.M., Zheng, T.X., Yang, Q.: Energy-efficient transmission design in non-orthogonal multiple access. IEEE Trans. Veh. Technol. **66**(3), 2852–2857 (2017)
12. David, H.A., Nagaraja, H.N.: Order Statistics, 3rd edn. Wiley, New York (2003)
13. Ding, Z., Poor, H.V.: Cooperative energy harvesting networks with spatially random users. IEEE Sig. Process. Lett. **20**(12), 1211–1214 (2013)
14. Gradshteyn, I.S., Ryzhik, I.M.: Table of Integrals, Series, and Products, 6th edn. Academic Press, New York (2000)

IoT Based Experimental Study to Modify Water Consumption Behavior of Domestic Users

Aneeq-ur Rehman[1], Reeba Raza[1], Naveed Ul Hassan[1], Yuren Zhou[2], Rui Liu[2], Benny Kai Kiat Ng[2], and Chau Yuen[2(✉)]

[1] Lahore University of Management Sciences (LUMS), Lahore 54792, Pakistan
{16100212,17060006,naveed.hassan}@lums.edu.pk
[2] Singapore University of Technology and Design (SUTD), 8 Somapah Road, Singapore 487372, Singapore
{yuren_zhou,rui_liu}@mymail.sutd.edu.sg, {benny_ng,yuenchau}@sutd.edu.sg

Abstract. In this paper, we present the results of a small experimental study to understand and quantify the impact of real time feedback on water consumption behavior inside buildings. We develop a low cost water monitoring node, which can be conveniently installed without plumbing requirements on water fixtures found in typical households and commercial buildings. In our experiment, we installed the developed Internet of Things (IoT) node on a kitchen sink in a commercial building. The sink was used primarily for washing mugs, plates, and making tea & coffee. We collected, analyzed and compared the data of different users for different activities (e.g., washing mugs and plates) to understand their water consumption pattern. Then, we provided real time feedback for three weeks to two major water consumers after every activity about potential water wastage. We observed a significant improvement in the water consumption behavior of these users (water wastage reduction up to 50%). This study clearly demonstrates the utility of low cost IoT based solutions and real time feedback in modifying water consumption behavior of domestic users.

Keywords: Internet of Things · Consumer behaviour · Feedback
Water conservation

1 Introduction

Water is perceived to be abundantly available and is provided at a nominal cost (often free of charge) to domestic consumers, but they do not consume it with due care and prudence. Empirical evidence suggests that even conservationist consumers, who claim to possess a positive attitude towards water management and climate change issues, also fail to reliably translate their noble intentions into concrete actions. This lack of prudence often leads to lot of water wastage. Unfortunately, many consumers remain unaware of the water wastage that results from their own actions [1].

© ICST Institute for Computer Sciences, Social Informatics and Telecommunications Engineering 2018
P. H. J. Chong et al. (Eds.): SmartGIFT 2018, LNICST 245, pp. 200–209, 2018.
https://doi.org/10.1007/978-3-319-94965-9_20

According to the psychological literature, human behavior can be influenced by the 'availability bias' [2,3]. In nutshell, availability bias suggests that consumers resort to mental shortcuts and draw on readily available information. Resources are not used with due care because often their conservation is not at the forefront of our consciousness. Consumers can be made to think and act more rationally if feedback and information is provided to them about their consumption habits and any wastage resulting from their actions. This, however, requires a better understanding of their existing resource consumption pattern, i.e., how, when and where the resources are being consumed by different users.

Internet of Things (IoT) platforms have gained tremendous popularity and research interest in recent years. However, in most buildings, sensing and control capabilities are very limited, and it is considerably hard to monitor real time resource utilization. A variety of IoT systems using wireless sensor networks for monitoring domestic usages of water have been proposed and developed in recent years [4–8]. In [4], a case study was presented in which a global household water consumption monitoring system was developed across two countries that showed remote, near real-time monitoring of water consumption in different households. Several systems have also been proposed to provide feedback and induce resource conservation behavior in the consumers [5–8]. [5,6] used android and web applications to show water consumption by users in terms of graphs. [7] developed a system that included automatic billing as a way of intervention to change consumer resource consumption pattern, while the system in [8] focused on providing real time alerts in the form of alarms when the consumer crossed some previously defined threshold on consumption.

In this paper, we also develop a low cost IoT based system to monitor and change the water consumption behavior of domestic consumers. Contrary to previous works, our paper disintegrates the water consumption data according to the activities performed by each user. In the proposed system model, the sensor nodes, which can be retrofitted on faucets and fixtures monitor the water consumption pattern of consumers. The data collected from the nodes is transmitted to a central gateway and then to a remote servers where the data is stored and processed to understand the consumer behavior. We performed an experiment where the system was deployed to monitor the water consumption pattern of kitchen sink users in a commercial building. The collected data was used to identify water wastage in different activities performed by two major water consumers in our study. These users were then provided real time feedback for three weeks after every activity about the potential water wastage. The results showed promising reduction in water wastage (up to 50%), which indicates the potential of low cost IoT platforms in modifying consumer behavior.

The rest of the paper is organized as follows. In Sect. 2 we describe an IoT based system model for water monitoring in buildings, in Sect. 3, we discuss our experimental water monitoring setup, in Sect. 4 we describe the results of our experimental study, while the paper is concluded in Sect. 5.

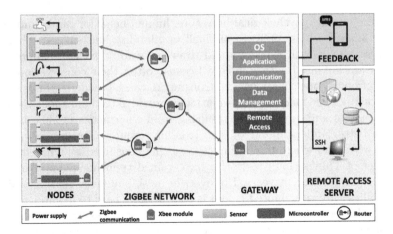

Fig. 1. IoT based system model for water monitoring

2   IoT Based System Model for Water Monitoring

An IoT based system model for water monitoring is shown in Fig. 1. There are four distinct components of the system. The sensor nodes monitor the water flow at different types of faucets and pass the data to a central gateway using a Zigbee network. The gateway aggregates and processes the data and sends it to a web server where the data is uploaded onto a database. The server can be remotely accessed to perform data analysis and the feedback is generated for the consumers. The gateway then communicates the feedback to the consumers. Further details of these modules are explained in the following subsections.

2.1   Water Monitoring Node

The most important module in this system is the node designed to fit on the faucet where the water usage needs to be monitored. To install a sensor, consideration of pipe size and its dimensions is of fundamental importance. Water pressure is also important for sensors that fit inline or somewhere in the piping infrastructure because high pressure can damage the sensor. Water pressure for domestic faucets and fixtures is at most 150Psi [9]. Flow rate of every faucet varies and the sensing module must have a reasonable measuring accuracy and sensitivity within that range. Moreover, the issues of additional plumbing requirements, such as, whether the sensor needs to be mounted inside the piping infrastructure vertically or horizontally and whether the sensor needs additional instrumentation for its deployment are also important.

The most commonly used sensors to monitor the water flow at the faucets are Hall-effect sensors [10]. These type of sensors can be easily interfaced with any microcontroller. As the water flows through the sensor, the magnetic rotor spins at a rate proportional to the flow of the water passing through the sensor. The rotor provides for the magnetic field to the sensor and a series of voltage pulses are generated whose frequency is proportional to the water flow rate.

2.2 Radio Communication Network

A low-energy radio, like Zigbee, could be used to transmit the data to a central gateway. Zigbee can form any network topology based on the number of nodes and the distances between the nodes. Zigbee has three device types: coordinator, router and end device [11]. Different routers should be placed by doing experimentation to determine the Received Signal Strengths (RSSI) of the Zigbees in Non-line of sight (NLOS) and Line of Sight (LOS) locations [12].

2.3 IoT Gateway

Gateway acts as an edge device which connects the sensor nodes to the internet. A gateway can play multiple roles as it collects the data coming in from all the sensor nodes, and then standardizes the formats of all the data so that it can be easily processed and stored. Some key features of the software stack of the gateway include operating system, application container, communication & connectivity, data management and remote management [13]. Raspberry Pi has all the key features and therefore it is often a popular choice to build a gateway in IoT based systems [14].

3   Experimental Low Cost Water Monitoring Setup

To test the performance of our water monitoring system and to study the impact of real time feedback in modifying consumer behavior, we built a low cost water monitoring system based on the model discussed in the previous section. We conducted an experiment by installing our water monitoring node on a kitchen sink in a commercial building. A significant amount of water is used at our targeted node for various activities; the most common being, washing dishes and mugs, making tea and coffee for the staff (almost 50 persons). The setup of our node is shown in Fig. 2.

Fig. 2. Experimental setup

3.1   Development of Water Sensing Node

We selected **YF-S201** Hall-effect sensor to develop our water sensing nodes [15]. The sensor does not have any plumbing or retrofitting requirements and it can be easily installed at the exterior of the faucet with minimal intrusions and obstructions. With proper calibration, this sensor also provides reasonable accuracy (85%–90%). Furthermore, it can measure water flow rate of up to 0.5 l/s, which falls in the range of the domestic faucets. This sensor can be fitted inline (both in horizontal and vertical mounting) or to the exterior of a faucet and piping infrastructure according to the application requirements. Moreover, this sensor is readily available in the market and can be purchased off the shelf.

A microcontroller is required to process the signals coming from this sensor. We used Arduino because it can be easily interfaced with both the sensor and the Zigbee modules [16]. Whenever the faucet turns on, an interrupt is generated and the voltage pulse produced by the sensor is detected by the Arduino. Arduino then disables the interrupt and calculates the flow rate and the total water consumption. The calculated data is then sent to the gateway using Zigbee network.

3.2   Zigbee Radio Network

Xbee is an embedded wireless communication module that is built on Zigbee standard [17]. To form a wireless network, we used Xbee S2C Pro models [11]. The Xbee at the gateway was configured as a coordinator and the one at the node was configured as an end device. The Xbees in between the coordinator and the end device were configured as routers. We placed two routers between the node and the gateway at distances where the RSSI value dropped below -70 dBm [12]. The exact router locations were determined by performing experiments for RSSI values in line of sight and non-line of sight paths.

3.3   Raspberry Pi Based Gateway

To build a gateway, we used Raspberry Pi 2, model B [14] as it supports all inbound and outbound communication protocols required for a typical IoT gateway. We used Xbee for the inbound communication, which is the data coming from the sensor node over the network formed and for the outbound communication we used Ethernet which connected our Raspberry Pi to the internet. The default operating system for the Raspberry Pi is Linux, specifically Raspbian, and it also runs Apache, as the web server; MySQL, as a database; and PHP, for server-side scripting [18]. Raspberry Pi is also capable of acting as a local web server [19]. The packets of flow rate, current flow, and total water consumed were separated and uploaded on the database table corresponding to the time-stamp using a python application. This gateway could be accessed through Secure Shell (SSH), which is a network protocol that provides administrators with a secure way to access a remote computer [20].

4   Experimental Study

In this section, we describe and discuss the results of our experimental study. The collected data was stored in a database along with the time-stamps. To associate water usage with individual consumers, we also attached a camera with the Raspberry pi and as soon as the faucet was turned on, the camera took a series of snapshots with time-stamps. Passive Wi-Fi tracking techniques [21] can also be used to distinguish the users using the faucet. In this method, the MAC addresses of the smart phones can be scanned and RSSI values can be used to find out the users near the faucet at the time of water consumption activity. This method will be further explored in future work.

The experiment was divided into two phases. In the first phase, we simply observed the activities and corresponding water consumption of all the users and logged them for five working days (1 week). We analyzed this data and selected the consumers who used the maximum amount of water during the experiment. In the second phase of the experiments, real time feedback was provided to the selected users and the impact of our feedback was observed in their water consumption behavior.

4.1   Phase I: Monitoring

The water consumption of all the users was monitored. The consumption patterns and disaggregation of data was done for only those users who contributed significantly or were involved in major activities at the node. The main activities observed were washing plates, mugs or hands.

Figure 3 shows the percentage of water consumed by different users. The three most frequent users are indicated as users X, Y and Z. Together, these users accounted for 66% of the total water consumption at the node. User X was found to be the highest water consumer, followed by user Y, and then user Z.

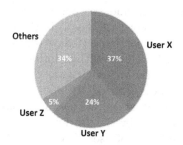

Fig. 3. Percentage water consumption by different users

We further analyzed the water consumption pattern of only the three main users. Table 1 shows their mean daily water consumption and variance. It can be observed that user X has the highest mean of 38.32 l. The mean daily water consumption by user Y is relatively less than user X but the variance is very high. High variance shows the inconsistency in the amount of water consumed by user Y for his activities. User Z has the lowest mean daily water consumption and is also relatively consistent. Therefore, for further analysis and feedback we also dropped user Z.

Table 1. Mean and variance of daily water consumption by users X, Y and Z

Users	Mean (*liters*)	Variance
X	38.32	29.90
Y	24.40	56.59
Z	4.89	1.56

The major daily activities of users X and Y were washing mugs and plates. However, the number of washed items varied for both the users. Due to this variation in the number of activities performed by the two users, we computed the average water consumption of washing a mug or a plate in that particular day. All the further analysis was done using the averaged value of the activity.

4.2 Phase II: Real-Time Feedback

The second phase of our experiment was aimed at providing feedback to users X and Y in our study. We initially briefed these consumers about their water usage pattern as observed in the monitoring phase and also reminded them about the importance of water conservation. Then after every activity that was performed by user X or Y, we notified them about the water (in liters) that was consumed. The feedback was provided for three consecutive weeks (5 days per week).

In Figs. 4 and 5 we plot the average water consumption of user X and user Y respectively to wash one mug and one plate for the complete experimentation period. The first 5 days (day 1 till day 5) in these two graphs indicate the pre-feedback period, while the remaining 15 days (day 6 till day 20) indicate the post-feedback period. It can be observed that during the pre-feedback week, the total water consumed to perform an activity was the highest for both the users. However, as the feedback process started, it can be observed that the water consumption of user X decreased drastically. The water consumption stayed almost constant in the second week and by the third week it began to increase slightly. On the other hand, the response of user Y to the feedback was very effective. As can be seen from Fig. 5, during the pre-feedback period, the consumption of water by user Y was very high, but during the feedback period the consumption rate significantly dropped. This indicates that the impact of feedback on user Y was relatively high as compared to user X.

To further understand the impact of feedback on consumer behavior, in Fig. 6, we plot the average weekly water consumption for both the activities by users

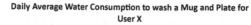
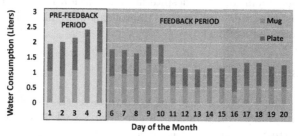

Fig. 4. Daily average water consumption by user X to wash 1 mug and 1 plate for the complete experimentation period

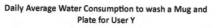
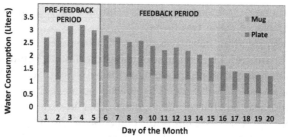

Fig. 5. Daily average water consumption by user Y to wash 1 mug and 1 plate for the complete experimentation period

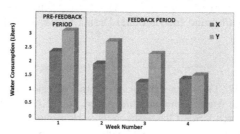

Fig. 6. Average weekly water consumption by users X and Y for washing mugs and plates

X and Y. From this figure, we deduce that the water consumption of user X was reduced by 42% from the pre-feedback week till the last week of the feedback. While user Y's water consumption was reduced by 53%. It can also be observed that the water consumption of user Y for the same set of activities was always higher than that of user X throughout the four weeks of our experiments. This also leads to the conclusion that despite large reduction in water consumption, user Y still wasted water and further improvement in his behavior could be achieved by continued feedback.

5   Conclusions and Future Work

The aim of this paper is to present our work towards the development of a low cost IoT based system, which could be effectively used to monitor the water consumption pattern of domestic consumers and then design consumer behavior change interventions. We successfully developed and deployed a water monitoring system for a kitchen sink node in a commercial building. With the help of the data collected by our system, we can easily identify the water consumption patterns of different users. In our experimental study, we identified two major water consumers and provided them real time feedback about potential water wastage. We observed tremendous positive impact of feedback on the water consumption pattern of these two users.

The impact of providing incentives along with feedback is an interesting area to be explored in future. The consumers using the faucet could be categorized as green and non-green users and a points based system could be devised to promote pro-conservation behavior. The overall points gained by various same faucet users could also be shared among each other in order to create a competition. Moreover, the water consumption behavior of the users once feedback is discontinued would also be evaluated in future.

Acknowledgment. This research is supported by the Higher Eduction Commission Pakistan through grant number SRG-425, LUMS Center for Water Informatics & Technology (WIT), National Research Foundation (NRF) Singapore and administered by Building and Construction Authority (BCA) - Green Buildings Innovation Cluster (GBIC) Programme Office and International Design Center (IDC) Singapore.

References

1. Miller, E., Buys, L.: The impact of social capital on residential-affecting behaviors in a drought-prone Australian community. Soc. Nat. Resour. **21**, 244–257 (2008)
2. Elisha, R.F., Karen, S., Elizabeth, V.H.: Household energy use: applying behavioural economics to understand consumer decision-making and behavior. Renew. Sustain. Energy Rev. **41**, 1385–1394 (2015)
3. Dale, S.: Heuristics and Biases: the science of decision making. Bus. Inf. Rev. **32**, 93–99 (2015)
4. Yang, S.H., Chen, X.M., Yang, L., Chao, B., Cao, J.: A case study of internet of things: a wireless household water consumption monitoring system. In: IEEE 2nd World Forum on Internet of Things, Milan, Italy, pp. 681–686 (2015)
5. Jamaluddin, A., Harjunowibowo, D., Rahardjo, D.T., Adhitama, E., Hadi, S.: Wireless water flow monitoring based on Android smartphone. In: International Conference of Industrial, Mechanical, Electrical, and Chemical Engineering, Yogyakarta, Indonesia, pp. 243–247 (2016)

6. Bodhe, A.R., Singh, R., Bawa, A.: An internet of things solution for sustainable domestic water consumption. In: International Conference on Computation System and Information Technology for Sustainable Solutions, Banglore, India, pp. 224–229 (2016)

7. Anjana, S., Sahana, M.N., Ankith, S., Natarajan, K., Shobha, K.R., Paventhan, A.: An IoT based 6LoWPAN enabled experiment for water management. In: 9th International Conference on Advanced Networks and Telecommuncations Systems, Kolkata, India, pp. 1–6 (2015)

8. Dutta, P., Dontiboyina, U.S.G.V.: Faucet add-on water supply management system using smart sensors. In: 2nd International Conference on Computational Intelligence Communication Technology, Ghaziabad, India, pp. 468–471 (2016)

9. Household Water Pressure, DIY by Example. http://www.diybyexample.info/2014/06/household-water-pressure

10. Sriratana, W., Murayama, R.: Application of hall effect sensor: a study on the influences of sensor placement. In: IEEE International Symposium on Industrial Electronics, Taipei, Taiwan, pp. 1–5 (2013)

11. DIGI: XBee/XBee-PRO® S2C Zigbee® RF Module (2017)

12. Mayalarp, V., Limpaswadpaisarn, N., Poombansao, T., Kittipiya, S.: Wireless mesh networking with XBee. In: 2nd ECTI-Conference on Application Research and Development, Chonburi, Thailand (2010)

13. The Three Software Stacks Required for IoT Architectures, Eclipse IoT White Paper (2016)

14. Raspberry Pi 2 Model B, Raspberry PI. https://www.raspberrypi.org/products/raspberry-pi-2-model-b

15. Rahul, R.I.: The water flow monitoring module. Int. J. Eng. Res. General Sci. 4(3), 106–113 (2016)

16. Kanishk, S., Manish, J., Madhur, M., Lokesh, Y., Nidhi, V.: Digital water meter using arduino. Int. J. Eng. Manag. Res. 7(2), 276–279 (2016)

17. ZigBee Wireless Standard, Digi International. http://www.digi.com/resources/standards-and-technologies/rfmodems/zigbee-wireless-standard

18. Kaur, S., Kaur, K., Singh, D.: A framework for hosting web services in cloud computing environment with high availability. In: IEEE International Conference in Engineering Education: Innovative Practices and Future Trends (AICERA), Kottayam, India, pp. 1–6 (2012)

19. Harrington, W.: Learning Raspbian. Packt Publishing, Birmingham-Mumbai (2015)

20. Barret, D.J.: PuTTY for windows. In: SSH, the Secure Shell: The Definitive Guide, 2nd edn. Reilly Media Inc., Newton (2005)

21. Scheuner, J., Mazlami, G., Schoni, D., et al.: Probr - a generic and passive WiFi tracking system. In: Proceedings of IEEE Conference on Local Computer Networks, Dubai, UAE (2016)

A Sustainable Marriage of Telcos
and Transp in the Era of Big Data:
Are We Ready?

Salman Naseer[1,4(✉)], William Liu[1], Nurul I. Sarkar[1], Peter Han Joo Chong[1],
Edmund Lai[1], Maode Ma[2], Rangarao Venkatesha Prasad[3], Tran Cong Danh[5],
Luca Chiaraviglio[6], Junaid Qadir[7], Yue Cao[8], Jinsong Wu[9],
Raymond Lutui[1,10], and Shahid Manzoor[4,11]

[1] Auckland University of Technology, Auckland, New Zealand
salman.naseer@aut.ac.nz
[2] Nanyang Technological University, Singapore, Singapore
[3] Delft University of Technology, Delft, The Netherlands
[4] University of Punjab, Lahore, Pakistan
[5] University of Science, Ho Chi Minh, Vietnam
[6] University of Rome Tor Vergata, Rome, Italy
[7] Information Technology University, Lahore, Pakistan
[8] Northumbria University, Newcastle upon Tyne, UK
[9] Universidad de Chile, Santiago, Chile
[10] Christ's University in Pacific, Sacramento, USA
[11] Swedish University of Agricultural Sciences, Uppsala, Sweden

Abstract. The emerging smart city paradigm e.g., intelligent transport,
smart grid and participatory sensing etc. is to advance the quality, performance and experience of urban citizten services through greater connectivity. This paradigm needs to collect data from citizens, various devices
and assets that could be monitored, processed and analysed for the city
governers to make better decision and also more efficiently manage those
assests and resources. While the telecommunication and Internet are
progressively being over-burdened and congested by the growing data
transmission demands. To keep expanding the telecommunications and
Internet infrastructures to accomodate these intensive data demands is
costly and also the associated energy consumptions and carbon emissions
could at long last wind up genuinely hurting the environment. To face
this issue in the coming era of big data, we envision it will be best to utilize the established urban transport and road infrastructure and existing
daily massive vehicular trips, to complement traditional option for data
transmission. After detailing the current state-of-the-art, we consider the
main challenges that need to be faced. Moreover, we define the main pillars to integrate the telecommunications and transport infrastructures,
and also a proposal for the future urban network architecture.

Keywords: Smart city · Intelligent transport · Energy consumption
Data offloading · D2D communication · Carbon footprints
Delay tolerant network

© ICST Institute for Computer Sciences, Social Informatics and Telecommunications Engineering 2018
P. H. J. Chong et al. (Eds.): SmartGIFT 2018, LNICST 245, pp. 210–219, 2018.
https://doi.org/10.1007/978-3-319-94965-9_21

1   Introduction

The emerging smart city (SC) paradigm needs to collect data from citizens, various devices and assets that could be monitored, processed and analysed for the city governers to make better decision and also more efficiently manage those assests and resources. Cisco predicts that a SC having a population of 1 million could generate 180 million GB data per day or 42.3 ZB/month [1]. One of the fundamental issues of SC is the accumulation of big data and information generated by different data sources remotely and to transfer this data to some predefined data centers. These data sources in the SC can be interconnected through various transmission media. However, the idea of "smart" leads basically the efficiently and wisely utilization of infrastructures and resources. These sources may include Smart Grid (SG) sensors, environmental monitoring systems, Smart health monitoring systems, field sensors, video surveillance stations. For example in SG, smart devices and meters are installed throughout the city for management, controll, and monitoring. These SG devices generate a tremendous amount of the data, which needs to be transmitted to utility control centers in order to manage the SG services. It is predicted that this SG data yields a 8000-fold increase in daily data volume, and in 2015 it is augmented to over 75200 TB [2]. Figure 1 [3] shows Cisco's forecast in 2017 for the mobile traffic Exabyte per month till 2021, and the projected data traffic almost sevenfold in 5 years' time. Global mobile data traffic grew 74% in 2015. It is 4000-fold vs. past 10 years, 400-million-fold vs. 15 years. The total Internet traffic has experienced a dramatic growth in the past 2 decades and is still growing very fast. Data rates of 5G are five times faster than the 4G whereas the mobile data growth is sevenfold in 2021 [4]. Hence data offloading is a promising solution to meet this exponential growth of mobile data as shown in Fig. 1.

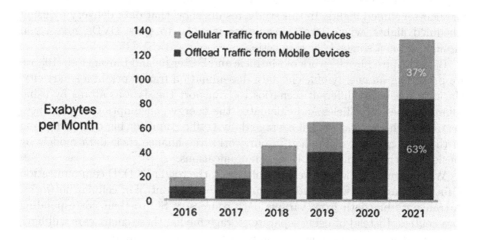

Fig. 1. Offloading needs of total mobile data traffic [3].

The rest of the paper is organized as follows. In Sect. 2 we review the relevant literature in delay tolerant network and D2D communications. In Sect. 3 our detailed vision is reported. In Sect. 4 we define our proposed network architecture in detail. Finally, the paper is concluded in Sect. 5.

2   Related Work

We address the issues of Delay Tolerant Network (DTN) [5] and particularly one of its variant, the Assisted Delay Tolerant Network (ADTN). In ADTN, we use different data carriers like cars, buses, trains, airplanes, etc. for store-carry-forwarding the data. These data carriers having radio interfaces, processing power and storage capacity act like moving routers and enhance the network capacity.

In [6], authors use throw boxes having radio communication, processing, and storage capabilities, to enhance the network capacity by deploying these boxes at planned positions in the network. Instead of using fixed nodes, some other approaches uses the moving nodes to increase the network throughput by adding some randomness through node mobility. In [7], random moving data MULEs (Mobile Ubiquitous LAN Extensions) are used to collect data opportunistically from data sources to data processing centers. Instead of using randomness, authors exploit the use of non-random movement of scheduled vehicles to transfer data from some selected sources on their pre-defined routes to the final destinations.

The above-defined literature enhances the network capacity in a small geographical area. Whereas in [8], to transport data on the large geographical area the authors propose the use of mobile gadgets of passengers, waiting for their scheduled flights. Data is loaded onto the mobile gadgets of the passengers with respect to their destinations, while they are sitting in the waiting hall to wait for their scheduled flights. In this study, results show that data delivery by using scheduled flights, when data transmission is equal to three DVDs, give equal throughput as a single TCP connection.

Being inspired by the work of Marincic and Foster [9] and Baron et al. [10], we are proposing an energy efficient data dissemination framework in a smart city. Marincic presented different scenarios to transport the data as atoms by using different types of vehicles and calculated the energy consumptions. Baron uses electronic vehicles and annual average daily traffic count for big data offloading on the road network of France. In our work, we enhance their data models by considering the advantage of D2D communications.

We efficiently utilize the vehicle volume on the road and D2D communication of future smart cars with roadside units, under the control of cellular networks, to transport big data from various data sources of SC to their corresponding data centers. Instead of using separate storage media, these smart cars will have their own storage, processing, and radio communication modules. These modules contain Wi-Fi, GPS, as well as cellular interfaces. In this paper, we propose a framework that offloads the data from congested networks and reduces the delay, energy cost, and carbon emissions through these smart vehicles by using their routine rides.

3 Our Vision

We first define the main pillars which, we believe, are essential for the design and management of proposed system, for delay tolrent data offloading in SC environment. In the following we sketch the proposed architecture.

3.1 Main Pillars

D2D Communication: To address the challenges of cellular data crises, a promising solution, device-to-device (D2D) communication has been suggested for next generation cellular networks. In D2D communication, mobile devices in wireless range can communicate with each other without traversing through the base station of cellular network or through the cellular backhaul. By exploiting the vicinity of mobile devices and direct communication, D2D data transmission can enhance the throughput, reduce the communication delay of mobile devices and it can enhance overall spectrum utilization, network throughput and performance [11]. Because of different advantages of D2D communication, proximate communication is appropriate for many user cases and it can introduced different peer-to-peer and location based services and applications.

A sound literature of D2D transmission is available, in this section we discuss the research issues of D2D communication and some existing solutions. In [12], multi-hop communication was proposed by using D2D communication in cellular networks. D2D transmission is analogous to ad hoc networks, the main difference is it often includes cellular network for controlling. A common classification of D2D communication is explained in literature [13]. With respect to spectrum utilization, D2D communication is divided into two strategies: (i) Inband D2D communication and (ii) Outband D2D communication. In first strategy, same energy spectrum band as of cellular communication is used for D2D communication, whereas in second strategy, to avoid the signal interference from cellular networks, some other bands like Wi-Fi, Blue Tooth, Wi-Fi Direct are used [14]. It is to be noted that, in outband D2D transmission, eNodeB BS can has different level of control over D2D transmissions such as connection establishment and neighbor discovery, this control can be managed by control channels of cellular networks. Otherwise D2D data transmissions work in self-governing mode. On the basis of spectrum resource utilization, Inband D2D data transmissions can be further characterized into two categories. If same spectrum resources as of cellular communications are used for D2D communication then it is called underlay D2D data transmission and if resources are reserved used for D2D communication from cellular spectrum band then it is called overlay D2D transmission [15].

Opportunistic Communication: When a number of users request for same contents in a given area, service provider governs the communication as an in charge and transferred the requested contents to the base station, and base station forwarded these contents to those users. These requested contents can be delivered to users by using three different approaches. In first case, the base station forward the requested contents directly to each user. In second case some of the target users are selected by the base station, it forward the data to the target users, and they disseminate the data among their peer users. Where as in third case, target users may store the requested contents and delivered to peer users when they are in their wireless range. If the selected users don't meet the timelines to transfer data, then cellular base station are responsible for deriving data to each of data-undelivered-subscriber. The latter case is referred as an opportunistic communication in [16]. Opportunistic communication also reduce the load on cellular base station that's why it can also be counted as a data offloading scheme. Target and peer user selection have main importance in opportunistic data offloading, these are explained here in more detail.

Target Carrier Selection: In content delivery networks there are several number of distributed server instead of a single server. These distributed servers store the more popular contents in their caches which are the most periodically requested contents by the users from main server. Now, it is not necessary for all users to get data from main server, they can get data from their nearby distributed server, hence the main server can remain uncongested. The same idea is implemented in opportunistic data communication in which target users cache the most popular contents in their cache and behave like distributed servers as in content delivery networks. In [17], a solution for target user selection is formulated and named as Mobile Social Network (MoSoNet). The users who already received the data from content delivering servers, or stored the contents that are requested by some other users, can be selected as a target user and can disseminate data to the other peer users who are selected as receivers in opportunistic social network.

In opportunistic networks, it is an important question that who is most suitable user as a target user among all users in network. In [18], one of the criteria is social importance of a user and it is defined as a number of peer users connected to that user. However, this not sufficient for selecting a user as a target user because it is possible that peer users are not interested in the contents that are stored by that users. Hence, another perimeter is defined as number of interested peer users in network for the selection of target user. Therefore, it is a trade-off between two these two factors for decision making over the selection of target user.

In another point of view, the problem of target user selection can also be formulated as an optimization problem having an objective of maximizing the number of peer users (number of content-delivered users) in a required time period so that load on the base station can be reduced. Different methods are defined in [17]. For an already defined number of selected users, in an approach of greedy algorithm, a set of target users are selected that maximizes the number of peer users or content-delivered users. As it can be considered that, greedy algorithm doesn't consider the location/time correlation and statistical consistency of user's daily presence. On the other hand in heuristic algorithms, this consistency is considered at an initial stage of target user selection. For example, consider a company whose employees remain present at its location at almost the same time during each working day of the week. In such a scenario, heuristic algorithm selects initial target user set same as before instead of defining the new set every day. Finally, initial target user selection set is picked up completely in a random way from all users. Simulation results of [17] showed that all these selection algorithm mitigated the traffic load on cellular network as compared to the scenario where data is delivered to each user by base stations themselves.

Peer User Selection: Peer users are those ones that accept the data contents directly from the selected target users. Data contents are transferred from base station to selected target users, and they can transfer these contents to the peer users in their wireless range via Bluetooth or Wi-Fi, by using an application like Haggle [18]. They can also disseminate the contents to their peers via some cellular resources in their social friend network like Twitter or Face book as mentioned before in Mobile Social Network (MoSoNet) [17].

D2D and Opportunistic Based Vehicular Networks: Due to the advantages of opportunistic and D2D technology, it is suitable for many vehicular related use cases, and can enable peer-to-peer and location based application and services. For instance, considering the enormous number of connected smart cars (connected cars in 2021 [1]), software update of smart car can put a noteworthy load on the cellular infrastructure, and cost a lot of money for car owners. Thus, the software update can be downloaded by some selected vehicles and then it can be transferred to other vehicles by opportunistic and D2D transmission. In this process, target vehicles can be selected by cellular network by applying some efficient algorithms and allocate appropriate resources to reduce the interference, optimize the performance and fulfill different QoS requirements. In this way, most of the cellular load and data traffic can be offloaded to V-D2D communication, and thus cost, energy and much of the cellular bandwidth can be saved. Furthermore, software update does not require the real time communication, it can compromise on some delays, in such cases the vehicular delay tolerant network (VDTN) can be employed to forward the software update package in a store-carry-forward manners which can further save the cost, energy and offload the cellular traffic. Audio/video streaming and gaming are some other type of data services among vehicular users such as social network in the vehicular proximity [19]. Generally, such types of applications are maintained by Wi-Fi direct, LTE-direct or DSRC communications, because of long device paring time and collisions they may not satisfy the proper requirements. V-D2D and opportunistic communication can support such type of services in a better way due to the use of cellular links for control channels. Cellular-control may provide connection setup within short delay time and batter resource allocation.

There are numerous research works that consider the problems of applying D2D communications in vehicular ad hoc networks. In [20], Cheng et al. investigate the validity of D2D communication for intelligent transportation systems (ITS) by applying spatial distribution of high mobility vehicles and on different channel characteristics. Simulation study showed that among all other D2D techniques, D2D-underlay got highest efficiency in spectrum utilization and transmission rate increases with decrease in distance of vehicles. Moreover, it is observed that with gradual increase in link density of V-D2D the average spectral efficiency first begins with growth then gradually declines, because when V-D2D link density increases then the network will get more severe interference. In [21], Sun et al. formulated a resource allocation strategy for cellular user equipment's (CUEs) and vehicular user equipment's (VUEs), to exploit the CUEs sum rate while assurance the strict delay reliability and consistency requirements of VUEs' services.

Business Insider (BI) intelligence research, predicts in 2021 for 82% of all cars will be shipped as connected cars [1]. These connected smart cars will have storage capacity, cellular, global positioning system (GPS), and wireless interfaces. If we consider Auckland city as a SC, it has 0.72 per capita vehicles in 2015 and this value is going to increase by each year as forecasted by the ministry of transport New Zealand. It predicts that the big volume of connected vehicles can produce a big bandwidth on the road according to our proposed system.

4   Proposed Network Architecture

We propose an opportunistic D2D based vehicular network data delivery architecture that utilizes vehicles (and also their existing trips), roadside networks, and D2D communications to assist to deliver those delay tolerantable data from different data sources to control centers. Data can be forwarded to and carried by vehicles and uploaded to data centers through SRU, instead of being transmitted through the cellular networks.

To demonstrate the key concepts, here we provide a smart-health service system model in Auckland city, New Zealand (Fig. 2) and its associated network architecture (Fig. 3). The cellular network coverage is ubiquitously available in the entire city and vehicles move around the city on various roads. The participating vehicles in this proposed architecture have been enabled with a dual mode, they can form an opportunistic network with roadside units and can be connected to the cellular network. The data forwarding mode selection algorithm will be placed in the central controller (CC) at the base station (BS). CC is to govern the overall communications between data sources and destinations via VDTN by using D2D communication. Whenever a data source wants to send data, it sends data request packet to CC by using cellular control signals. CC decides the most suitable option to transfer data via Internet or via VDTN by using the information of delay tolerant indicator (associated with data demand), available vehicle volume on the road, vehicle's trajectory history, and minimum energy cost. On the basis of these optimization results, CC decides, if core network is more suitable for the given request then the CC directs the source to

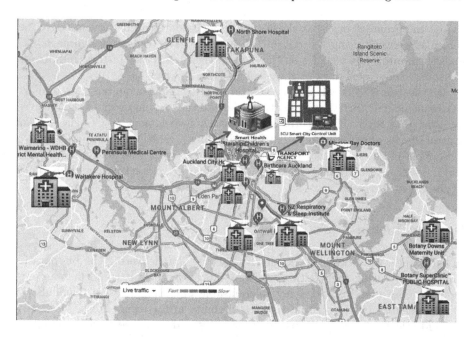

Fig. 2. Proposed system overview.

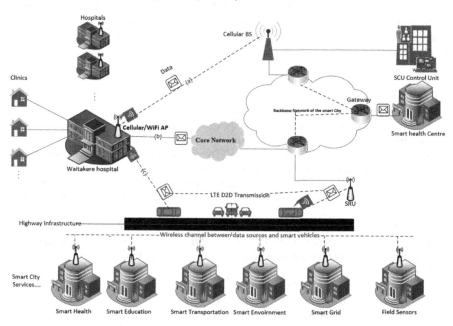

Fig. 3. Proposed system overview: data aggregation from various data sources to the central control units for storage and analysis.

choose an option (a) or (b) as shown in Fig. 3, else data can be forwarded by the vehicular network using the option (c). For this data transfer assignment, CC finds the appropriate vehicles on the basis of their trajectory history and directs the source to forward data to these vehicles one by one.

On the receiver side whenever a vehicle carrying data encounters with a smart city roadside unit (SRU), it uploads the data to the local network of the receiver by using that SRU. The receiver sends an acknowledgment to CC via cellular control signals after receiving a data bundle from a vehicle. In this paper we are more focused on the overall network architecture, while the analytical models and numerical validation on this proposed network architecture could be found in our recent work [1].

5   Conclusion

In this paper, we propose a network architecture for delivering the big data in smart city by using the existing transport infrastructure. It could best accommodate those massive and delay tolerant-able data transmission demands by utilizing efficiently the existing vehicles' mobility in the urban areas. Moreover, it could release the congestions in the wireless and wired networks, as well as reduce the energy consumption and carbon emissions. The automatic selection algorithm (i.e., selection of end-to-end data delivery path among wired, wireless or vehicular networks) for optimal data dissemination is to be discussed. In addition, the more complex mobility models and numerical validation studies are also needed in the future work.

References

1. Naseer, S., Liu, W., Sarkar, N.I., Chong, P.H.J., Lai, E., Venkatesha Parsad, R.: A sustainable vehicular based energy efficient data dissemination approach. In: Conference Proceedings: 27th International Telecommunication Networks and Applications Conference (ITNAC), Melbourne, November 2017
2. Yu, R., Zhang, Y., Gjessing, S., Yuen, C., Xie, S., Guizani, M.: Cognitive radio based hierarchical communications infrastructure for smart grid. IEEE Netw. **25**(5) (2011)
3. Cisco Visual Networking Index: Global Mobile Data Traffic Forecast Update, 2016–2021 White Paper, CISCO, August 2017. http://www.cisco.com/c/en/us/solutions/collateral/service-provider/visual-networking-index-vni/mobile-white-paper-c11-520862.html
4. Dean, J.: 4G vs 5G mobile technology. December 2014. https://www.raconteur.net/technology/4g-vs-5g-mobile-technology
5. Cao, Y., Sun, Z.: Routing in delay/disruption tolerant networks: a taxonomy, survey and challenges. IEEE Commun. Surv. Tutor. **15**(2), 654–677 (2013)
6. Zhao, W., Chen, Y., Ammar, M., Corner, M., Levine, B., Zegura, E.: Capacity enhancement using throwboxes in DTNs. In: 2006 IEEE International Conference on Mobile Adhoc and Sensor Systems (MASS), pp. 31–40. IEEE (2006)

7. Shah, R.C., Roy, S., Jain, S., Brunette, W.: Data mules: modeling and analysis of a three-tier architecture for sparse sensor networks. Ad Hoc Netw. **1**(2), 215–233 (2003)

8. Keränen, A., Ott, J.: DTN over aerial carriers. In: Proceedings of the 4th ACM Workshop on Challenged Networks, pp. 67–76. ACM (2009)

9. Marincic, I., Foster, I.: Energy-efficient data transfer: bits vs. atoms. In: 24th International Conference on Software, Telecommunications and Computer Networks (SoftCOM), pp. 1–6. IEEE (2016)

10. Baron, B., Spathis, P., Rivano, H., de Amorim, M.D.: Vehicles as big data carriers: road map space reduction and efficient data assignment. In: IEEE 80th Vehicular Technology Conference (VTC2014-Fall), pp. 1–5, September 2014

11. Golrezaei, N., Mansourifard, P., Molisch, A.F., Dimakis, A.G.: Base-station assisted device-to-device communications for high-throughput wireless video networks. IEEE Trans. Wirel. Commun. **13**(7), 3665–3676 (2014)

12. Lin, Y.-D., Hsu, Y.-C.: Multihop cellular: a new architecture for wireless communications. In: Proceedings of the Nineteenth Annual Joint Conference of the IEEE Computer and Communications Societies, INFOCOM 2000, vol. 3, pp. 1273–1282. IEEE (2000)

13. Asadi, A., Wang, Q., Mancuso, V.: A survey on device-to-device communication in cellular networks. IEEE Commun. Surv. Tutor. **16**(4), 1801–1819 (2014)

14. Asadi, A., Mancuso, V.: On the compound impact of opportunistic scheduling and D2D communications in cellular networks. In: Proceedings of the 16th ACM International Conference on Modeling, Analysis Simulation of Wireless and Mobile Systems, pp. 279–288. ACM (2013)

15. Pei, Y., Liang, Y.-C.: Resource allocation for device-to-device communications overlaying two-way cellular networks. IEEE Trans. Wirel. Commun. **12**(7), 3611–3621 (2013)

16. Dimatteo, S., Hui, P., Han, B., Li, V.O.: Cellular traffic offloading through WiFi networks. In: 2011 IEEE 8th International Conference on Mobile Adhoc and Sensor Systems (MASS) pp. 192–201. IEEE (2011)

17. Han, B., Hui, P., Kumar, V.A., Marathe, M.V., Shao, J., Srinivasan, A.: Mobile data offloading through opportunistic communications and social participation. IEEE Trans. Mob. Comput. **11**(5), 821–834 (2012)

18. Peng, W., Li, F., Zou, X., Wu, J.: The virtue of patience: offloading topical cellular content through opportunistic links. In: 2013 IEEE 10th International Conference on Mobile Ad-Hoc and Sensor Systems (MASS), pp. 402–410. IEEE (2013)

19. Luan, T.H., Shen, X., Bai, F., Sun, L.: Feel bored? Join verse! engineering vehicular proximity social networks. IEEE Trans. Veh. Technol. **64**(3), 1120–1131 (2015)

20. Cheng, X., Yang, L., Shen, X.: D2D for intelligent transportation systems: a feasibility study. IEEE Trans. Intell. Transp. Syst. **16**(4), 1784–1793 (2015)

21. Sun, W., Ström, E.G., Brännström, F., Sui, Y., Sou, K.C.: D2D-based V2V communications with latency and reliability constraints. In: Globecom Workshops (GC Wkshps), pp. 1414–1419. IEEE (2014)

Review of Cost Optimization of Electricity Supply by Using HOMER and a Case Study for a Big Commercial Customer in Brazilian Amazon Area

Abdulrahim Hamed S. Alghamdi,
Carlos Henrique Marciano Rodrigues Castro, and Ramon Zamora[✉]

Electrical and Electronic Engineering Department,
Auckland University of Technology, Auckland, New Zealand
{wgs0484,mtr0836}@autuni.ac.nz, ramon.zamora@aut.ac.nz

Abstract. Renewable energy utilization for electricity supply has increased significantly. Technology maturity, cost reduction, and environmental friendliness are significant factors that encourage this increase. The expansion of distributed generation is increasing significantly due to the concern of many householders and entrepreneurs in minimizing the energy costs at residential and commercial properties due to the high significance of the energy costs at total expenses of most families and business. This study focuses on exploring optimization process of energy costs of a grid-connected hybrid PhotoVoltaic (PV)/Battery/Grid. The profile of a big commercial load located in Brazilian Amazon area is used as an optimization example. The system performance and optimization results are verified by using Hybrid Optimization of Multiple Energy Resources (HOMER).

Keywords: HOMER software · Energy management
Energy cost optimization

1 Introduction

Solar systems applications in smart grids and hybrid systems are recognized all over the world as a financially and technologically viable solution for electricity supply. Solar applications have increased 53% per year in the US over the last five years, and 60% per year globally [1]. The growth of distributed generation, mainly leaded by the number of installed solar systems, results in great challenge and more complexity for traditional grids and utilities [2]. Smart grid systems has an important role to support the growth of renewable energy sources, promising reliability, security, resiliency and efficiency for the electric grid [3].

Residential, commercial and industrial applications of solar energy are set to increase even more once the technology is already well established. While solar energy is more environmentally friendly and sustainable, as well as promising for remote areas, investment in a solar system is dependent on the financial viability compared to other energy sources. Solar energy price is governed by the costs of equipment and

© ICST Institute for Computer Sciences, Social Informatics and Telecommunications Engineering 2018
P. H. J. Chong et al. (Eds.): SmartGIFT 2018, LNICST 245, pp. 220–229, 2018.
https://doi.org/10.1007/978-3-319-94965-9_22

installation. The equipment price links to the technical specifications; the higher the capability, the higher the cost. Consequently, the equipment cost will have a significant impact on the decision to develop a hybrid system using solar energy. By investing in hybrid systems, commercial businesses can control their energy prices and even freeze their energy costs for future years at current prices. After all, such businesses as supermarkets have free space on their roofs to install large solar panels and energy storage systems.

To guarantee minimal investment costs to supply a specific load profile, optimization methods are required to effectively design hybrid systems on smart grids. HOMER is an optimization method capable to model and simulate a considerable number of system combinations, in order to achieve the best technical and economical results for the studied system.

This study considers an economic and technical approach of optimization model for hybrid solar energy for commercial businesses in order to reduce the cost of energy. By proposing an optimal energy cost, entrepreneurs have reason to invest in hybrid renewable energy. This paper will demonstrate the benefits of HOMER software usage in power generation and optimization of energy systems. A literature review of how HOMER helped in optimizing different systems to produce a better system in terms of power generation and economic aspects is also included. As a study case, the load profile of a big commercial load in Brazilian Amazon area is analysed for a PV/Battery/Grid system by using HOMER.

2 Overview of HOMER Optimization

In this section, HOMER software will be introduced. Then, different type of power systems from several case studies in different countries will be reviewed. The case studies are stand-alone PV system and hybrid renewable energy systems with and without grid connection. The typical standalone system consists of one of Renewable Energy Sources (RES) and involves a storage system, and the hybrid system contains either more than one RES or include a conventional source of energy and storage system with or without grid connection.

2.1 HOMER Software

The Hybrid Optimization of Multiple Energy Resources (HOMER) microgrid software was developed by the U.S. National Renewable Energy Laboratory (NERL) to aid in designing microgrid systems as well as to ease comparing different sources of power generation for a wide-range applications [4, 5]. In addition, HOMER assists in building techno-economic and reliable systems which can include renewable and non-renewable energy sources, storage systems and load management [4]. It is used by more than 120,000 users in more than 190 countries. HOMER facilitates the users to compare many different design options in regards to their technical and economic aspects [5]. Moreover, it helps in considering any changes in the inputs of the modelled systems. HOMER provides users with the ability to include off-grid and grid-connected systems to supply different loads, and also contains various combination of wind turbine,

PhotoVoltaic (PV), small hydro, biomass power, batteries, fuel cells and hydrogen storage.

There is a challenge of designing and analysing microgrid systems because of a number of available options and the uncertain key parameters. Renewable energy sources add the complexity due to their intermittency and availability. HOMER was designed to overcome these difficulties. HOMER performs three tasks which are simulation, optimization and sensitivity analysis.

In the simulation task, HOMER test the performance of the designed systems each hour of the year in order to find its technical feasibility and life-cycle cost. In the next stage (optimization), HOMER searches for the best fit system from many different simulations of different systems considering the constraints given by the users at the lowest life-cycle cost. Lastly, the sensitivity analysis in HOMER does consider multiple optimizations considering any change in the inputs to have a better understanding of the effects of any uncertainty in the modelled systems.

2.2 Review of Standalone Solar Energy Systems

A standalone PhotoVoltaic (PV) system consists of PV module, a storage system, controller, inverter and a load as shown in Fig. 1. This standalone solar energy system has been increasingly used in developed and developing countries [6]. A review of a case study with various options of single renewable source and hybrid systems is discussed in [7]. The load tested in that study is 50 rural households (24.4 MWh/year), and HOMER software was used to assess the systems. The standalone system consisting of PV/Battery is selected as the most cost-effective system for the location (Iran). The analysis results show the Cost of Energy (COE) of $0.247/kWh–without allowing power supply shortage–and Net Present Cost (NPC) of $120,738.00. In another study in India, different power generation combinations were tested and found that 150 W PV/60 Ah Battery system is the most economical system for 219 kWh/year load [8]. The COE is $0.258/kWh without subsidy. With subsidy, it becomes

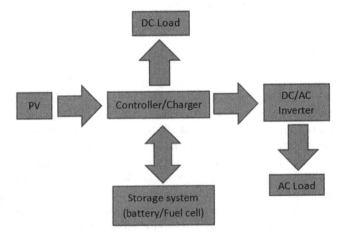

Fig. 1. The general structure of standalone PV system [6].

$0.145/kWh and the system is more economic. With subsidy by 50%, the solar system is more viable at distance greater than 6.08 km as the grid system costs more. Nevertheless, without the subsidy, the solar system is still more cost effective than the grid when the distance is 8.64 km or more. In a rural area in south of Iraq, a health clinic load of 11.534 MWh/year was analysed by HOMER [9]. In this study, a standalone 6 kW PV, 3 kW inverter and 80 batteries (225 Ah and 6 V) was selected as the most economic system with COE of 0.238 US$/kWh.

2.3   Review of Hybrid Renewable Energy Systems Involving Solar Energy

A Hybrid Renewable Energy System (HRES) consists of more than one RES with or without Conventional Energy Sources (CES), grid connection, controller, converter/inverter and a storage system. The general structure of HRES involving solar energy (PV) is shown in Fig. 2.

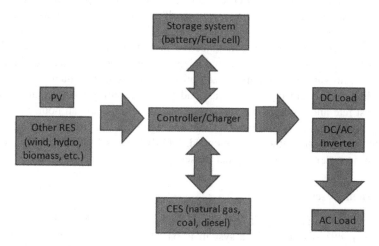

Fig. 2. The general structure of HRES involving solar energy (PV) [10].

In India, AC home appliances were used as a load for power generation analysis with the energy consumption of 1.825 MWh/year [10]. The output of this analysis was that the most cost-effective system is the hybrid system of PV, wind turbine and Vision 6FM 200D with the values of 2 kW, 1 kW and 12 V, 200 Ah battery, respectively, with COE of $1.232/kWh. Another example of HRES with CES involvement, a 2 kW PV, 4 kW diesel generator and 2 Surrette 6CS25P battery (6 V, 1156 Ah) were selected to be the optimal solution for the load of 17.52 MWh/year in remote area of Jordan [11]. The COE of that system was $0.297/kWh. In Saudi Arabia, a study analysis were done using HOMER for six different systems consisting of solar, wind, converter, battery banks, fuel cells, electrolyser and hydrogen tank for different cities and some chosen international cities [12]. The analysis results in the most economical hybrid system involving RES which consists of PV array, wind turbine, converter and battery storage bank with the values of 2 kW, 3 kW, 2 kW and 7 banks respectively for Yanbu city in Saudi Arabia. The COE was $0.609/kWh. This study was done for a typical house with an average load demand of 14 kWh/day.

3   Optimization Model Using HOMER

To demonstrate the usage and effectiveness of HOMER optimization method, it will be used to analyse a load profile of a commercial building located in Brazil.

In hybrid systems, the optimal size of each system component is a complex task and involves several variables. The optimal size of hybrid systems will lead to a good relation between performance and cost. It has been shown that optimally designed hybrid systems have been found to be cost effective and reliable [13].

As optimization model, a load profile of a Brazilian commercial building was used in this research. The hybrid system is modelled as a PV/battery/grid connected system. The optimization process is focused on identifying and proposing the optimal PV/battery system size that simultaneously result in lower grid energy dependency and lower energy prices.

In this system, the optimization objective is to minimize the Levelized Cost of Energy (LCOE). To evaluate the optimal size for the proposed system, different configurations from various component size options of the solar installation are analysed. After adding the system inputs, such as load profile, monthly solar radiation, annual real interest rate and a range of equipment sizes, HOMER starts an hourly simulation for every possible system configuration, computing the power from the grid, PV array and batteries. Once the simulation is ended, HOMER sorts feasible combinations for the system in order to increase the net present cost and minimize the cost of energy.

3.1   System Description

The studied system is located in the city of Tucuruí, state of Pará – Brazil, located at 3°45'53.0"S, 49°40'16.0"W. The system is based on a commercial property with a public of 2.000 customers per day on average. The load is basically composed of motors and compressors from many refrigerators, as well as lighting. Additionally, the daily operation hours is Monday – Sunday from 8 am to 8 pm.

The system characteristics are detailed bellow:

- Daily consumption: 2,550.23 kWh/day
- Monthly consumption: 79,057.13 kWh/month
- Grid Energy Price: $0.28/kWh
- Power Factor: 0.87
- Maximum demand: 180.43 kW
- Roof top area: 3,600.00 m^2

3.1.1   Load Profile and Solar Irradiation

The load profile in Fig. 3 shows the daily load profile of the studied system. The daily average demand is 2,550.23 kWh/day with a peak demand of 180.01 kW. The monthly average solar Global Horizontal Irradiance (GHI) data for the location of 3°45'53.0"S, 49°40'16.0"W is shown in Fig. 4. The annual average solar GHI is 5.01 kWh/m²/day.

According to the graphs above, the load profile presents a high correlation with solar irradiation hours. As a result, the load profile will also correlate with solar energy production, which represents a great opportunity for economical savings.

Fig. 3. System daily load profile.

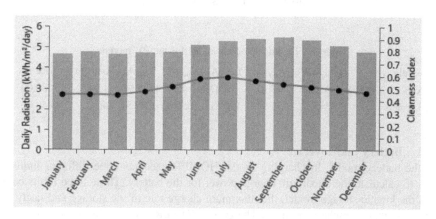

Fig. 4. Tucuruí solar irradiation.

3.1.2   PV Panels

The PV panels generate DC electricity from the sunlight exposure. The panels tilted at an angle of $0°$ due North with slope of $3.75°$ and the derating factor is 80%. The PV panels have a capital and replacement costs of 888.889 $/kW with lifetime of 20 years for the system. This price is of a JA Solar 315 W Silver Poly Pallet consisting of 23 PV modules [14]. The technical specifications for this PV panels are shown in Table 1.

The output power of the PV array is expressed by (1) based on [15].

$$P_{PV} = Y_{PV} f_{PV} \left[\frac{\bar{G}_T}{\bar{G}_{T,STC}} \right] \tag{1}$$

where:

Y_{PV} is the rated capacity of the PV array under Standard Test Conditions (STC) [kW],

f_{PV} is the PV derating factor [%],

$\bar{G}_T$ is the solar radiation incident on the PV array in the current time step [kW/m^2],

$\bar{G}_{T,STC}$ is the incident radiation at STC [1 kW/m^2].

The above equation neglects the temperature effect. However, if the effect of the temperature is considered, then the above equation is modified as in (2).

$$P_{PV} = Y_{PV} f_{PV} \left[\frac{\bar{G}_T}{\bar{G}_{T,STC}} \right] \left[1 + \alpha_P \left(T_c - T_{c,STC} \right) \right] \tag{2}$$

where:

α_P is the temperature coefficient of power [%/$^\circ$C],

T_c is the PV cell temperature in the current time step [$^\circ$C],

$T_{c,STC}$ is the PV cell temperature under STC [25 $^\circ$C].

Table 1. Technical specifications for the PV module.

Module type	P_{mp} (W)	V_{mp} (V)	I_{mp} (A)	V_{oc} (V)	I_{sc} (A)	Area (m^2)	Price ($)
JAP6-72-315/4BB	315	37.19	8.47	45.95	8.98	1.94	280.00
23 panels	7245	855.37	194.81	1056.85	206.54	44.58	6440.00

3.1.3 Battery and Inverter

For the battery model, in charging mode, HOMER executes three different limits in order to calculate the maximum charge power for the battery. These three limits come from the kinetic storage model, the maximum charge rate of the storage and lastly the storage's maximum charge current. Therefore, HOMER equates the maximum storage charge power to the least of these three values of the previous limits with the assumption of each applies after charging losses and is shown in (3).

$$P_{batt,cmax} = \frac{MIN \left(P_{batt,cmax,kbm}, P_{batt,cmax,mcr}, P_{batt,cmax,mcc} \right)}{\eta_{batt,c}} \tag{3}$$

where:

$P_{batt,cmax,kbm}$ is the maximum amount of power that can be absorbed by the two-tank system (kinetic storage model),

$P_{batt,cmax,mcr}$ is the maximum amount of the storage charge power regarding the maximum charge rate of the storage,

$P_{batt,cmax,mcc}$ is the maximum amount of storage charge power regarding the maximum charge current of the storage.

Table 2. Technical specification for the battery.

Model	Nominal capacity (Ah)	Voltage (V)	Minimum charge (%)	Price ($)
CR-430 (16#)	860	48	20	5544.00

All these three different maximum powers are calculated based on different equations and can be found in [15]. The technical specifications for the battery are shown in Table 2 [16]. This price is for 16 batteries of this model.

As the PV panels produces DC electricity, an inverter is required in order to supply AC loads. The selected inverter for this system has a capital and replacement cost of 73.573 $/kW with minimal lifetime of 12 and up to 25 years, as well as efficiency of 98%.

4   Results and Discussion

For this commercial building, a grid-connected PV system with battery banks was simulated to find whether or not the system is feasible to lower the LCOE. The modelled system is shown in Fig. 5.

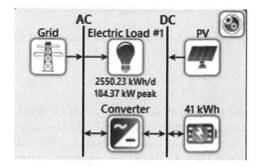

Fig. 5. Grid-connected PV system.

4.1   Optimization Results

As can be seen from Fig. 6, the optimization results in having 463 kW of PV array, 100 kW converter and 1600 batteries. The dispatch strategy that was used for this system is called load following (LF) which means that when the generator operates, it produces only enough power to meet the primary electrical load for the system.

Therefore, other loads and lower priority loads such as charging batteries and deferrable loads are left to the RES [15].

In Fig. 7, the costs of each component of the system and the total costs are shown. The total initial capital cost is $973,312.91, NPC is $2,506,631.00 and LCOE is $0.2004/kWh. Nevertheless, the high cost is due to the number of batteries used in the system. Even though the battery costs a lot and increases the overall cost of the system, the LCOE of the system is still lower than the COE of the grid price which is $0.28/kWh.

Architecture							Cost				System
⚠ 🖥 🖧 ⬆ ☒	PV (kW)	41 kWh	Grid (kW)	Converter (kW)	Dispatch	COE ($)	NPC ($)	Operating cost ($)	Initial capital ($)	Ren Frac (%)
🖥 🖧 ⬆ ☒	463	1,600	999,999	100	LF	$0.200	$2.51M	$118,609	$973,313	63

Fig. 6. Optimization result of the system.

Component	Capital ($)	Replacement ($)	O&M ($)	Fuel ($)	Salvage ($)	Total ($)
CR-430	$554,400.00	$0.00	$206,840.25	$0.00	$0.00	$761,240.25
Grid	$0.00	$0.00	$1,262,667.69	$0.00	$0.00	$1,262,667.69
JA Solar 315W PV	$411,555.61	$0.00	$59,854.43	$0.00	$0.00	$471,410.04
System Converter	$7,357.30	$5,571.58	$0.00	$0.00	($1,615.62)	$11,313.26
System	$973,312.91	$5,571.58	$1,529,362.37	$0.00	($1,615.62)	$2,506,631.24

Total NPC:	$2,506,631.00
Levelized COE:	$0.2004
Operating Cost:	$118,608.90

Fig. 7. The cost of each component and overall cost of the system.

5 Conclusion

This paper has provided a review of cost optimization of electricity supply by using HOMER. It discussed some examples of PV only standalone system and hybrid systems where HOMER was the tools used for analysing the technical and economic feasibilities of the systems. Then, HOMER was applied to optimize electricity supply of a big commercial building in Brazil. The optimal solution was a grid-connected PV system that produced 463 kW from PV arrays. Since the PV arrays were installed on the roof top, the installed PV size as well as the PV output power depend on the building roof area.

References

1. Torani, K., Rausser, G., Zilberman, D.: Innovation subsidies versus consumer subsidies: a real options analysis of solar energy. Energy Policy **92**, 255–269 (2016)
2. Tuballa, M.L., Abundo, M.L.: A review of the development of Smart Grid technologies. Renew. Sustain. Energy Rev. **59**, 710–725 (2016)
3. Bhalshankar,S.S., Thorat, D.C.S.: Integration of smart grid with renewable energy for energy demand management: Puducherry case study. Presented at the International Conference on Signal Processing, Communication, Power and Embedded System (SCOPES)-2016, Paralakhemundi, Odisha, India (2016)
4. HOMER Energy LLC. https://www.homerenergy.com/index.html
5. Lambert, T., Gilman, P., Lilienthal, P.: Micropower system modeling with HOMER. In: Integration of Alternative Sources of Energy, pp. 379–418 (2006)
6. Oko, C., Diemuodeke, E., Omunakwe, N., Nnamdi, E.: Design and economic analysis of a photovoltaic system-a case study. Int. J. Renew. Energy Dev. **1**, 65 (2012)
7. Askari, I.B., Ameri, M.: Techno-economic feasibility analysis of stand-alone renewable energy systems (PV/bat, Wind/bat and Hybrid PV/wind/bat) in Kerman, Iran. Energy Sour. Part B: Econ. Planning Policy **7**, 45–60 (2012)
8. Kamalapur, G., Udaykumar, R.: Electrification in rural areas of India and consideration of SHS. In: 2010 International Conference on Industrial and Information Systems (ICIIS), pp. 596–601 (2010)
9. Al-Karaghouli, A., Kazmerski, L.: Optimization and life-cycle cost of health clinic PV system for a rural area in southern Iraq using HOMER software. Sol. Energy **84**, 710–714 (2010)
10. Dalwadi, P., Shrinet, V., Mehta, C., Shah, P.: Optimization of solar-wind hybrid system for distributed generation. In: 2011 Nirma University International Conference on Engineering (NUiCONE), pp. 1–4 (2011)
11. Nema, P., Dutta, S.: Feasibility study of 1 MW standalone hybrid energy system: for technical institutes. Low Carbon Econ. **3**, 63 (2012)
12. Al-Sharafi, A., Sahin, A.Z., Ayar, T., Yilbas, B.S.: Techno-economic analysis and optimization of solar and wind energy systems for power generation and hydrogen production in Saudi Arabia. Renew. Sustain. Energy Rev. **69**, 33–49 (2017)
13. Maleki, A., Ameri, M., Keynia, F.: Scrutiny of multifarious particle swarm optimization for finding the optimal size of a PV/wind/battery hybrid system. Renew. Energy **80**, 552–563 (2015)
14. JA Solar 315 W Silver Poly Pallet (23) of Solar Panels. https://www.wholesalesolar.com/1890060/ja-solar/solar-panels/ja-solar-315w-silver-poly-pallet-23-of-solar-panels
15. HOMER Help Manual. https://www.homerenergy.com/products/pro/docs/3.11/index.html
16. Crown 860AH 48VDC 41,280 Wh (16) Battery Bank. https://www.wholesalesolar.com/cms/crown-cr430-flooded-battery-specs-2834660653.pdf

Efficient Fault Identification Protocol for Dynamic Topology Networks Using Network Coding

Hazim Jarrah[1], Peter H. J. Chong[1(✉)], Nurul I. Sarkar[2], and Jairo Gutierrez[2]

[1] The Department of Electrical and Electronic Engineering,
Auckland University of Technology, Auckland, New Zealand
{Hjarrah, peter.chong}@aut.ac.nz
[2] The Department of Information Technology and Software Engineering,
Auckland University of Technology, Auckland, New Zealand
{nurul.sarkar, jairo.gutierrez}@aut.ac.nz

Abstract. This paper considers the problem of fault identification in dynamic topology networks using the time-free comparison model. Here, we introduce an efficient self-diagnosis protocol that can identify faulty nodes in dynamic networks. This protocol can correctly diagnose various fault types including permanent, dynamic, and soft faults. The protocol consists of a testing stage and a disseminating stage. During the testing stage, each node identifies the state of a part of nodes using the time-free comparison model. Afterward, nodes share their views employing a random linear network coding (RLNC) technique in the disseminating stage. The design of the disseminating stage is crucial for diagnosis efficiency. Using RLNC obviates the need for disseminating the views individually, and hence it reduces the number of messages required to diagnose the network. The OMNeT++ simulation has been used to evaluate the performance of the proposed protocol regarding the communication complexity. Results show that the proposed protocol is robust, scalable and energy-efficient.

Keywords: Self-diagnosis · Dynamic networks · Dynamic fault
RLNC

1 Introduction

Faults are the origins of impairments of network dependability [1]. That is, they hinder reliance on services provided by the network. However, they are inevitable and may evolve into errors and failures [2, 3]. In dynamic networks that have been deployed and operated in critical situations, faults hit more often as a result of unpredictable circumstances, and that may cause risks for people, environment and finance [4]. The development of dependable networks requires dealing with faults. Traditionally, system-level fault diagnosis problem [5] considers automated fault identification in wire and wireless networks [6–9]. Recently, we have developed a time-free fault diagnostic model that respects the design requirements of dynamic networks [10]. In particular, it takes into consideration the asynchronous communications and the topology changes in this kind of networks.

© ICST Institute for Computer Sciences, Social Informatics and Telecommunications Engineering 2018
P. H. J. Chong et al. (Eds.): SmartGIFT 2018, LNICST 245, pp. 230–239, 2018.
https://doi.org/10.1007/978-3-319-94965-9_23

The time-free comparison model is so named because it removes time constraints imposed by earlier models. Therefore, it suits asynchronous systems such as dynamic networks. Its design obviates the need for timers that are extremely hard to set into dynamic networks. Moreover, it is robust to dynamic topology changes. It is based on comparison approach where nodes examine each other states and collaborate to identify faulty and fault-free nodes in the network. In comparison approach, Nodes execute a task appointed, and their outputs are compared to identify whether any node is faulty. The main idea behind the comparison approach is that fault-free nodes processing the same inputs (task) will produce the same outputs.

This paper introduces a distributed self-diagnosis protocol to solve the diagnosis problem in dynamic networks. That is, this protocol enables fault-free nodes to determine the status of each node in the network correctly. Our proposed protocol consists of two main stages namely comparison stage and disseminating stage. The comparison stage employs the time-free comparison protocol to identify faulty nodes. This stage is executed by each node to test its neighbour nodes and generates a partial view of them. The disseminating stage exploits random linear network coding (RLNC) technique to exchange the partial views among nodes and generate a global view about the network. Using RLNC aims to reduce the number of diagnosis messages and hence enhance the protocol efficiency.

Network coding (NC) is an emerging communication paradigm introduced in 2000 [11]. This revolutionary paradigm remodels how nodes communicate with each other to improve the performance of networks. Nodes traditionally store and forward each packet received whereas they combine packets and forward the combination at once in NC [12]. To date, NC has been employed in numerous applications for wire and wireless networks [13]. The studies demonstrated that NC benefits network operation and design regarding throughput, scalability, robustness, energy, and reliability [14, 15]. Different coding techniques have been proposed. Nonetheless, random linear network coding (RLNC) considers more realistic assumptions for mobile networks [16]. In RLNC, Three operations are crucial; namely encoding, recoding, and decoding. Where source nodes encode native packets, intermediate nodes recodes coded packets, and destination nodes decode the packets. Firstly, source nodes generate an encoded packet that embodies linear combinations of native packets over a finite field using random coding coefficients. Intermediate nodes, then, recode these packets creating a linear combination of encoded packets. In this sense, these nodes can encode packets even though they have not been decoded yet. Finally, destination nodes decode coded packets and regenerate native packets once they get a sufficient number of linearly independent encoded packets.

The paper presentation includes sections as follows. Section 2 describes the system, the fault, and the diagnostic model. Section 3 presents a time-free fault diagnosis protocol using RLNC. Section 4 shows the simulation results obtained along with analysis. Finally, Sect. 5 concludes the paper.

2 Preliminaries

2.1 System and Fault Model

This subsection describes dynamic system considered in this research. The system consists of n mobile nodes that communicate wirelessly via packet radio network. It is an asynchronous system that alleviates time restrictions on node speed, transmission delay, and computation time. Nodes have no access to global clock. The dynamic network is represented by a communication graph that has dynamic topology and hence the connections may change over the time. At specific time t, a graph $G_t = (V_t, E_t)$ characterizes the network; V is the set of nodes and E is the set of edges; $E \subseteq V \times V$. A graph $G' = (V', E')$ indicates the fault-free nodes in the graph G at time t. The graph G' is assumed to comply with Assumption 1 which is an essential rule to preserve the properties of fault self-diagnosis protocols, i.e., correctness and completeness.

Assumption 1. Connectivity over Time: Let $G' \subseteq G$ be a subgraph that consist of fault-free nodes in G at time t. Then, there must be at least one path between every two nodes $u, v \in G'$. That is, $\forall u, v \in V', u \rightarrow v$.

We assume nodes are mobile. Thus, the neighbour nodes may change. Also, we assume that nodes adopt the passive mobility model where nodes are unaware of moving. Hence, they cannot inform neighbours about that. As a consequence, the neighbour nodes cannot distinguish between migrated out or undergoing a fault node.

Faults have been investigated from different perspectives. In particular, a fault could be soft or hard considering its impact on node communications. That is, hard fault (e.g., fail-stop, fail silence, and crash) prevents node communications while soft fault may impact node operations but not communications. Another perspective of fault is based on its duration. Permanent faults endure until be fixed by external interventions such as battery depleted or node crash. On the other hand, temporary faults, i.e. intermittent and transient faults, cease to exist spontaneously. Fault occurrence time has also been used to distinguish between static and dynamic faults. Static faults exist before the start of a diagnosis session whereas dynamic faults emerge during the diagnosis session.

This research assumes that links experience no faults. Mainly, no message creation, alteration or loss may happen through these links. Nodes, however, may undergo any type of faults except temporary faults.

System diagnosability is an upper limit to the number of faults that a system can undoubtedly diagnose. That is, if the number of faulty nodes exceeds that limit, then the system will be disconnected, and it is unsure to diagnose faults correctly and completely. Here, we utilise a local fault model that bounds the upper limit of faults locally. Assume σ_v is the upper limit of faulty nodes in v's neighbourhood. The σ_v is limited by the degree of the node v, $\deg(v)$, i.e., $\deg(v) > \sigma_v$. In case of reliable broadcast the bound, should be, $\deg(v) > 2\sigma_v$. This is because, it is uncertain to achieve reliable broadcast if half or more of nodes are faulty [17, 18].

Definition 1. Local Diagnosability: A dynamic network is locally σ-diagnosable at node v if each fault-free node can unambiguously identify the fault status of all nodes given that the number of neighbour faulty nodes does not exceed σ_v.

Fault-free nodes should be able to reply to $\sigma + 1$ test requests within the first α replies. This assumption implies that every fault-free node will be correctly diagnosed by at least one fault-free node. That is, fault-free nodes are winning nodes and achieve Assumption 2.

Assumption 2. Winning Nodes: Every fault-free node, u has a set of best neighbours that can communicate with u faster than with the other nodes.

2.2 Time-Free Diagnostic Model

The time-free diagnostic model has been developed for dynamic networks [10]. It adopts the comparison approach to identify faulty nodes. In the comparison approach [19], a task is appointed to nodes, and their outputs are compared to detect the state of nodes examined. This model assumes tasks are complete, i.e. they can detect faults in nodes. Also, it assumes that fault-free nodes executing the same task continually generate exact outputs while faulty nodes generate different outputs. Fault-free nodes can compare the outputs and generate comparison outcomes. This model utilises the asymmetric invalidation model rules [20] shown in Table 1.

Table 1. Comparison outcomes for the gMM model [20]

Nodes under comparison	Node performing the comparison	
	Fault-free	Faulty
None is faulty	0	0 or 1
One is faulty	1	0 or 1
Both are faulty	1	0 or 1

The following describes the steps to perform the comparisons in the time-free diagnostic model:

- Test Request Generation: A node u generates a test task T_i, where i is an integer number depicting the test number. Next, it broadcasts the test request message $m = (TEST, T_i)$, where $TEST$ indicates the message type. Afterwards, u waits for responses from αu nodes. It is noteworthy that u operates no timers.
- Test Request Reception: Once a node v receives the test request message m from u, it produces the result R_u^v of the task T_i. Then it broadcasts the test response message $= (RESPONSE, T_i, R_u^v)$, $RESPONSE$ is the message type. After that, the node v starts its diagnosis session through generating its test request message. As we consider a dynamic topology, v could be just moved into u's transmission range, i.e., $v \notin N_u$ at the test generation time.
- Test Response Reception: Consider a node $w \in V$. Upon receiving responses from αw nodes, w stops waiting. Then, w takes either of the two following actions:
  - Case 1: w knows the expected result of the task T_i; it compares them. If $R = R_u^w$, then w can conclude that v is fault-free, and hence, v will be added with an associated timestamp to the list of fault-free nodes diagnosed by w, i.e.,

$FF_w = FF_w \cup \{v, ct\}$. Otherwise, v is added to the list of faulty nodes with a timestamp ct, i.e., $F_w = F_w \cup \{v, ct\}$.

- Case 2: w does not know the expected result of the task T_i. Hence, w executes the task T_i first and then compares the result with R_u^v. If the comparison outcome is 0 then, it will add v, with a timestamp ct, to the fault-free list, $FFw = FFw \cup \{v, ct\}$. Otherwise, v will be added to the faulty nodes list, $Fw = Fw \cup \{v, ct\}$.

It is clear that this model obviates the need for timers and alleviates time constraints. Moreover, it tolerates the topology changes. Seemingly, fault-free nodes may improperly be diagnosed as faulty just because they have not replied fast. However, the correct state of any node is held with the highest timestamp by at least one fault-free node. Besides, nodes collaborate to identify a complete and correct view of the network. Therefore, in the end, the objective of having a complete and correct diagnosis of a network is guaranteed.

3   Network Coding-Based Self-diagnosis Protocol

This section introduces a novel distributed self-diagnosis protocol for dynamic topology networks considering static and dynamic permanent faults. Nodes collaborate with each other to diagnose the whole network in a distributed fashion. Every fault-free node executing this protocol obtains a complete and correct view of the faulty status of all the nodes in the network. The diagnosis session includes two main stages namely testing and disseminating. In the testing stage, nodes diagnose their neighbours creating partial views about the network. Next, they exchange these views with each other in the disseminating stage creating a global view.

The diagnosis session could be started either at fixed intervals or when a node detects an unusual event. A node then initiates the testing step. The messages that are transmitted during the diagnosis session are called diagnosis messages. The end of the session is when all nodes stop running the protocol.

The primary design objective of our protocol is to reduce the number of diagnosis messages. In particular, the design of the disseminating stage affects the protocol efficiency regarding communication complexity. Reducing the number of diagnosis messages is crucial for real deployment. In particular, the lesser the diagnosis messages during the diagnosis session, the more scalable and energy-efficient protocol. Also, the protocol design should be robust to topology changes. Considering these design requirements, we propose an RLNC-based distributed self-diagnosis protocol (RLNC-DSDP) for dynamic networks.

In our proposed protocol, the testing stage is based on the time-free diagnostic model described briefly in Sect. 2. Besides, the disseminating stage employs RLNC technique to improve the protocol efficiency. This protocol is executed on every node in the network. In the following, we detail these stages at a node, u.

1. Testing Stage

The testing stage at node u commences by sending a message of type *TEST* to neighbour nodes. This message includes a test task, T_i. That is, u sends a message $m = (TEST, T_i)$ to every one-hop neighbour nodes at that time $v \in N_u$. It is clear that neighbour nodes may change over the time as a result of network dynamics. Once a node v receives a diagnosis message for the first time, it performs the following procedures. First, it generates a message of type *TEST* including its test task if it have not started yet; $m_v = (TEST, T_i)$, and broadcast m_v to its neighbours; N_v. Second, it sends back a message of type *RESPONSE* including the task received (T_i) and the results calculated (R_u^v); $m = (RESPONSE, T_i, R_u^v)$. The response messages may be received by non-testers. However, including the test task in the response message helps other nodes to diagnose its state. Once u receives α messages of type *RESPONSE*, it generates a partial view based on the time-free comparison protocol in Sect. 2. That is, nodes will be diagnosed as fault-free if they produce the same results, soft-fault if they produce different results, and hard-fault if they send no reply. The node partial view contains two lists, namely fault-free list, FFu and Faulty list, Fu. These lists consists of members of form $(ID,)$, ID and ct represent node identifier and current timestamp respectively.

The design of this protocol uses message exchange pattern and obviates the need for timers. The result is a practical protocol for dynamic networks that is robust to topology changes. By the end of this stage, u maintain a partial view about adjacent nodes. Mobile or slow nodes may erroneously consider as faulty because they move away from u or they did not reply within the first α node. However, the system and diagnostic model assumptions guarantee that the correct state of any node is held by at least one fault-free node with the highest timestamp.

2. Disseminating Stage

At the end of the comparison phase, each node has a partial view of the network. The dissemination phase considers conveying these views to other nodes over the network to gain a complete view of all the nodes in the system. Our proposed protocol employs RLNC to perform this task. In the following, we describe how RLNC improves the efficiency during this stage.

Each node during this stage plays mainly two roles. First, it creates and transmits its partial view about the network. Second, it updates its view upon receiving other partial views and relays them to other nodes. In this sense, every node, u has an information message named *PartialView* that consist of the lists of faulty and fault-free nodes at u; $p_u = (PartialView, Fu, FFu)$. Hence, there are a set of messages to be exchanged among all nodes in the network; $\{p1, p2, p3, \ldots, pn\}$. Our proposed RLNC-based dissemination proceeds as the following. First, each node, u transmits its message p_u. Upon collecting α dissemination messages, u generates a coded packet, e, combining linearly packets received. That is, e is the total of multiplying each packet with the corresponding value in the coding vector, c as in the following relation:

$$e = \sum_{i=1}^{n} c_i P_i \qquad\qquad (1)$$

Where $c = (c1, c2, \ldots, cn)$ is a coding vector that composes coefficients selected randomly from a finite field, F. The node u generates and sends a message of type *ENCODED* containing the information vector along with the coefficient vector; $e_u = (ENCODED, DataV, CoeffV)$. During this session, u will add to its decoding matrix any message of type *ENCODED* that increase the rank of the decoding matrix. Also, this message will be forwarded to other nodes. However, the message received will be discarded if it has no innovative packet. Later, full ranked decoding matrix will be solved and the native messages will be retrieved, by Gaussian elimination. Hence, u has the partial views of all nodes except nodes experiencing a hard fault as they cannot communicate with other nodes. Therefore, it can generate a complete view about the system considering the most recent information.

It is clear that RLNC implements NC in a distributed fashion and it requires no earlier awareness about packets received by other nodes. However, RLNC adds computational overhead for nodes and transmission overhead attaching coefficient vectors to messages. These additional overheads may hinder RLNC uses under some scenarios.

4   Simulation Results and Analysis

This section presents the simulation results obtained under different scenarios to evaluate the performance of our proposed protocol. We have used OMNeT++ simulator to conduct our simulations.

First, we evaluate the protocol efficiency and scalability with regard to network's size. That is, various networks with different sizes varies from $10 - 100$ nodes. The nodes distributed randomly into the simulation area and 10% of nodes were considered faulty. The network topology is fixed in this scenario. The area size is $600\,m \times 600\,m$ and the transmission range is $150\,m$ for each node. Figure 1 shows the results under this scenario. The number of messages exchanged increases with the increasing number of nodes. However, it is clear that using RLNC significantly reduces the number of required diagnosis messages.

Second, we evaluated the protocol performance and robustness for increasing the number of faults. In this scenario, a network of 80 nodes had experienced 2 to 30 faults. The network had fixed topology and constant connectivity. Both static and dynamic faults have been considered in this scenario. Figure 2 depicts the results obtained under this scenario. It compares the communication complexity of our protocol experiencing static faults (RLNC-DSDP (Static)) as well as dynamic faults (RLNC-DSDP (dynamic)). It is clear that the increase in the number of faults leads to a decrease in the number of messages. The reason is that fewer number of nodes got involved in the diagnosis session. This figure also shows that our protocol is robust for various type of faults. That is, the diagnosis messages required for identifying dynamic and static faults are very close.

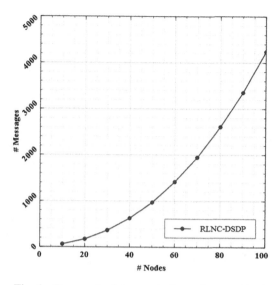

Fig. 1. Communication complexity under scenario 1

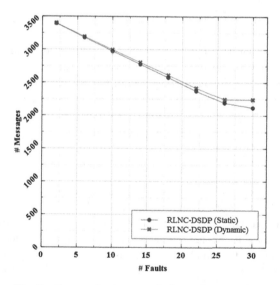

Fig. 2. Communication complexity under scenario 2

Finally, we evaluated the protocol performance under dynamic topology network. A network of 50 nodes has mobile nodes ranging from 2 – 10. The mobility of nodes caused random topology changes. Static and dynamic faults also considered under this scenario. Figure 3 shows the figures obtained under this scenario. This figure compares the performance under static and dynamic faults as well. The number of messages exchanged is reduced when more mobile nodes are there. It means that the mobility of

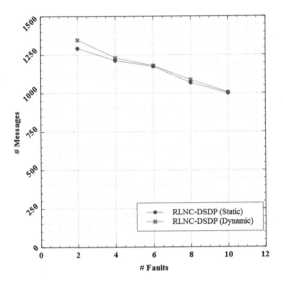

Fig. 3. Communication complexity under scenario 3

nodes aids in distributing the diagnosis messages. Also, it shows the robustness under various fault types.

The results show that the RLNC-DSDP protocol can efficiently identify the various type of faults in static and dynamic topology networks. That is, it provides a scalable, robust, and energy-efficient diagnosis algorithm. Further research is needed to investigate the computation overhead and delay caused by using RLNC. However, we believe that the advantages gained outperform the potential overhead.

5   Conclusion

In this paper, we proposed a new self-diagnosis protocol based on the time-free comparison model. Our protocol implements a simple RLNC technique to share the partial views. The partial views are grouped together instead of broadcasting them individually. Therefore, it obviates the need for broadcasting every local view received individually and hence the number of diagnosis messages required is significantly reduced. The future work concentrates on temporary faults identification.

References

1. Basile, C., Killijian, M.-O., Powell, D.: A survey of dependability issues in mobile wireless networks (2003)
2. Santoro, N.: Design and Analysis of Distributed Algorithms, vol. 56. Wiley, Hoboken (2006)
3. Jarrah, H., Sarkar, N.I., Gutierrez, J.: Comparison-based system-level fault diagnosis protocols for mobile ad-hoc networks: a survey. J. Netw. Comput. Appl. **60**, 68–81 (2016)

4. Silva, I., Leandro, R., Macedo, D., Guedes, L.A.: A dependability evaluation tool for the Internet of Things. Comput. Electr. Eng. **39**, 2005–2018 (2013)
5. Preparata, F.P., Metze, G., Chien, R.T.: On the connection assignment problem of diagnosable systems. IEEE Trans. Electron. Comput. **EC-16**, 848–854 (1967)
6. Blough, D.M., Brown, H.W.: The broadcast comparison model for on-line fault diagnosis in multicomputer systems: theory and implementation. IEEE Trans. Comput. **48**, 470–493 (1999)
7. Chessa, S., Santi, P.: Comparison-based system-level fault diagnosis in ad hoc networks. In: 2001 Proceedings of 20th IEEE Symposium on Reliable distributed systems, pp. 257–266 (2001)
8. Elhadef, M., Boukerche, A., Elkadiki, H.: Diagnosing mobile ad-hoc networks: two distributed comparison-based self-diagnosis protocols. In: Proceedings of the 4th ACM International Workshop on Mobility Management and Wireless Access, pp. 18–27 (2006)
9. Elhadef, M., Boukerche, A., Elkadiki, H.: Performance analysis of a distributed comparison-based self-diagnosis protocol for wireless ad-hoc networks. In: Proceedings of the 9th ACM International Symposium on Modeling Analysis and Simulation of Wireless and Mobile Systems, pp. 165–172 (2006)
10. Jarrah, H., Chong, P., Sarkar, N.I., Gutierrez, J.: A time-free comparison-based system-level fault diagnostic model for highly dynamic networks. In: Proceedings of the 11th International Conference on Queueing Theory and Network Applications, p. 12 (2016)
11. Ahlswede, R., Cai, N., Li, S.-Y., Yeung, R.W.: Network information flow. IEEE Trans. Inf. Theor. **46**, 1204–1216 (2000)
12. Ho, T., Lun, D.: Network Coding: An Introduction. Cambridge University Press, Cambridge (2008)
13. Deb, S., Effros, M., Ho, T., Karger, D.R., Koetter, R., Lun, D.S., et al.: Network coding for wireless applications: a brief tutorial (2005)
14. Matsuda, T., Noguchi, T., Takine, T.: Survey of network coding and its applications. IEICE Trans. Commun. **94**, 698–717 (2011)
15. Bassoli, R., Marques, H., Rodriguez, J., Shum, K.W., Tafazolli, R.: Network coding theory: a survey. IEEE Commun. Surv. Tutorials **15**, 1950–1978 (2013)
16. Chou, P.A., Wu, Y., Jain, K.: Practical network coding. In: Proceedings of the Annual Allerton Conference on Communication Control and Computing, pp. 40–49 (2003)
17. Koo, C.-Y.: Broadcast in radio networks tolerating byzantine adversarial behavior. In: Proceedings of the Twenty-Third Annual ACM Symposium on Principles of Distributed Computing, pp. 275–282 (2004)
18. Bhandari, V., Vaidya, N.H.: Reliable broadcast in radio networks with locally bounded failures. IEEE Trans. Parallel Distrib. Syst. **21**, 801–811 (2010)
19. Maeng, J., Malek, M.: A comparison connection assignment for self-diagnosis of multiprocessor systems (1981)
20. Sengupta, A., Dahbura, A.T.: On self-diagnosable multiprocessor systems: diagnosis by the comparison approach. IEEE Trans. Comput. **41**, 1386–1396 (1992)

Performance Evaluation of Handover Protocols in Software Defined Networking Environment

Dong-Ru Lee[1], Shan Jaffry[2], Syed Faraz Hasan[2], Yaw-Wen Kuo[1], and Xiang Gui[2(✉)]

[1] Electrical Engineering, National Chi Nan University, Puli, Taiwan
ywkuo@ncnu.edu.tw
[2] School of Engineering and Advanced Technology, Massey University, Palmerston North, New Zealand
{S.jaffry,F.hasan,X.gui}@massey.ac.nz

Abstract. Smooth handovers of Mobile Nodes (MN) between different points of attachments to the network ensures seamless connectivity when users move from one place to another. In IP based wireless networks, Mobile IPv6 (MIPv6) protocol manages handover issues, which has evolved with time to take many different forms. In this paper, we explore the softwarization of wireless networks with special regard to mobility management. We modify the legacy mobility management protocols to accommodate Software Defined Networking (SDN) features and measure their performance using simulations. The considered performance indicators include time to complete the handover process and the number of steps required for the same.

Keywords: Mobile IPv6 · Software Defined Networking · Handover Mobility

1 Introduction

Global Internet data transfer is largely enabled by a protocol called Internet Protocol (IP). The identifiers used by this protocol are called IP addresses. A mobile node (MN) will have to change its IP address as it changes the point of connection to the network. In order to enable seamless connectivity for MNs, a new protocol called Mobile IP (MIP) was proposed by Internet Engineering Task Force (IETF) [1]. With the introduction of IPv6 on the network layer, the mobile versions of the protocol also evolved as MIPv6 [2].

In a typical wireless network, a MN is connected to an Access Point (AP) over the radio interface. This AP is the first Point of Attachment (PoA) for the MN, which also acts as its gateway to the Internet. The APs are typically meant for indoor environments and have limited communication range (typically 200–300 m). Multiple APs are linked together to extend the range of a wireless

© ICST Institute for Computer Sciences, Social Informatics and Telecommunications Engineering 2018
P. H. J. Chong et al. (Eds.): SmartGIFT 2018, LNICST 245, pp. 240–249, 2018.
https://doi.org/10.1007/978-3-319-94965-9_24

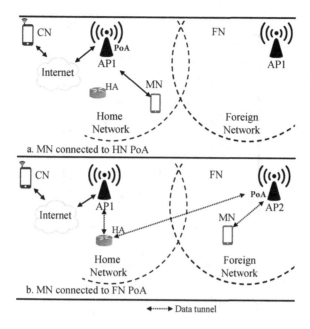

a. MN connected to HN PoA

b. MN connected to FN PoA

◀┄┄┄┄▶ Data tunnel

Fig. 1. An example of MIPv6 packet delivery in HN and FN

network. Due to mobility, a MN may change its PoA while still maintaining the connection via the next PoA. This process is called handover (see Fig. 1).

While multiple solutions have been proposed for handover management, they all lack adaptability due to the legacy hardwired infrastructure. However, the recent advances in Software Defined Networking (SDN) have provided opportunities for adaptable network deployment [3–5]. For example, the authors in [6] have shown that SDN can be programmed to support IP mobility, along with re-routing shortest path between a MN and its intended destination called the Corresponding Node (CN). The authors have shown that the hop-counts required in handover and the associated signaling overhead can be reduced using SDN. On the other hand, the authors in [3] have used SDN to realize intra-technology handover. In a previous work, the authors have proposed an SDN based implementation for Hierarchical MIPv6 (HMIPv6) protocol in [7]. Basu et al. have presented a framework for software enabled mobility management in [8] by distributing main controller's functionalities to a set of multiple sub-controllers. The distributed architecture is aimed to reduce signaling overhead in making local and global handover decisions at different controllers.

In this paper, we evaluate the performance of different variants of the legacy MIPv6 protocol with SDN. The performance parameters considered are signaling steps and handover latency. This paper is organized as follows. Section 2 discusses mobility protocols. Section 3 explains SDN and its underlying functions. SDN-enabled simulation is presented in Sect. 4 and conclusions are presented in Sect. 5.

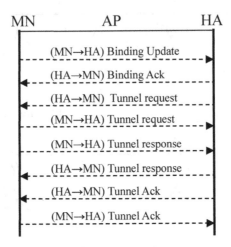

Fig. 2. An example of MIPv6 packet delivery in HN and FN

2   Handover Protocols

MIPv6 is the legacy protocol that handles handovers for MNs. Multiple variants of MIPv6 protocols have been proposed by IETF over the past years. Two of the most popular MIPv6 variants are Hierarchical MIPv6 (HMIPv6) and Proxy MIPv6 (PMIPv6).

2.1   Mobile IPv6

According to MIPv6, a MN connected to its Home Network (HN) is assigned an IP address called the Home Address (HoA), as shown in Fig. 1a. A MIPv6 based network must have at least one entity called Home Agent (HA) that stores HoA of the MN. A MN is always identified by its HoA when in home network. When a MN travels from its HN to another network, called Foreign Network (FN), it must performs handover to seamlessly change the PoA. While within FN, a MN is assigned a Care of Address (CoA). The MN must send its current CoA to HA. HA then binds the HoA to the CoA to form a data tunnel between MN and CN. The handover process completes when the MN's CoA is updated in HA. During handover, HA buffers all the incoming traffic for MN. The data tunnel enables data exchange from CN to the CoA of the MN, and back. As shown in Fig. 1b, all IPv6 packets destined for the MN's HoA are routed to its CoA by HA. Note that CoA may change but HoA remains the same until the session ends. The process of building tunnel is shown in Fig. 2. Note that at least one HA must be configured on the home network, and MN must know the IP address of HA.

2.2   Proximity Mobile IPv6

PMIPv6 is a network-controlled variant of MIPv6 in which, contrary to the legacy method, handover process is initiated by the network [9]. The initiation of the handover depends on parameters such as signal strength, etc. Hence, PMIPv6 offloads handover decision from MN to the network. This preserves the battery power for MN that is critical for its operation. The main entity in PMIPv6-based architecture is the Local Mobility Anchor (LMA), which basically acts as a HA. Each LMA manages the set of Media Access Gateways (MAGs). Each MAG is further responsible for managing and reporting the connection of MN with LMA. Most of the mobility management signals flow between MAG and LMA over wired links. The wired connectivity between LMA and MAG makes control signaling faster and more reliable [10].

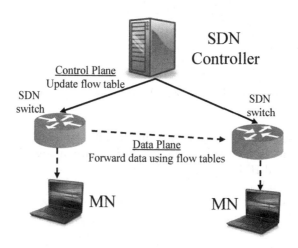

Fig. 3. A depiction of Software Defined Networking

2.3 Hierarchical Mobile IPv6

MIPv6 manages global (inter-network) and local (intra-network) handovers in the same way. HA may be located at multi-hop distances from the MN or it may even be in another network, which incurs additional signaling every time MN sends binding updates. This overhead grows larger in situations where local handover rate is high. In order to treat the local and global handovers differently (so to incur less overhead in the former), IETF has proposed Hierarchical MIPv6 (HMIPv6) [11]. HMIPv6 introduced a new node, called Mobility Anchor Point (MAP), which manages local handovers. On the other hand, the global handovers occur in same manner as in MIPv6. Instead of a single CoA, HMIPv6 assigns a Local CoA (LCoA) and Regional CoA (RCoA). For a handover within the same network, LCoA is changed whereas RCoA remains the same. The new LCoA is updated in the MAP though binding update process. Note that in HMIPv6, each AP is connected to the MAP. A set of APs and their MAP constitute a MAP mobility domain. When a MN travels between two different networks (i.e. two different MAP), both LCoA and RCoA change. The layering in HMIPv6 reduces

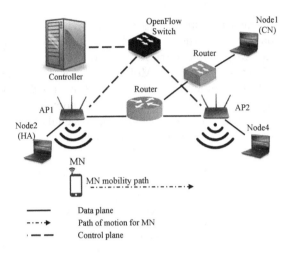

Fig. 4. SDN-enabled MIPv6 implementation

the redundant signaling, especially in high mobility environments because local and global mobility is managed separately. We have also modified legacy HMIPv6 as Network-Assisted HMIPv6 (NA-HMIPv6).

To this end, we have briefly discussed MIPv6 and its two variants that are relevant to this work. In the next section, we explain SDN so that its implementation for mobile IP can be explained in Sect. 4.

3   Software Defined Networking

SDN transforms a conventional hard-wired network into a programmable network by decoupling the control and data planes. For example, legacy routers forward incoming packets from one port to an out-going port based on routing table information. Hence, a router can be viewed as having two planes: (i) decision making plane (control plane), and (ii) packet forwarding plane (data plane). SDN decouples these two planes, providing more flexibility to control routing during run time, as shown in Fig. 3. This way, SDN manages to keep the network behaviour adaptive and dynamic.

The communication link between the SDN controller and its routers carries the information that updates the flow tables. For this purpose, several communication protocols have been proposed. OpenFlow has been the most popular choice [12].

3.1   OpenFlow Based SDN

OpenFlow (OF) is a communication interface between control and data planes of SDN. OF, which is also referred to as an OF-Switch, enables direct control of the routers' data plane via the controller. An OF-Switch maintains flow table(s),

Table 1. OF controller flow table (packet (P_n) of size N for flow k)

Packets	Actions
$P_{k,0}$	Re-direct to controller
$P_{k,i}$	Next hop to R_2 if $k = 0$, $\forall i \in N$
$P_{k,i}$	Next hop to R_3 if $k \neq 0$, $\forall i \in N$

which are used to control packet flow. The flow-table entries in an SDN-enabled router can be modified (i.e. additions, deletion of flow entries) by the OF controller in run time. Basic entries in a flow table are 'flows' and 'actions'. The flow-entry extracts packet information (e.g. MAC address, or incoming port etc.). The action table keeps information about the actions associated with the flow-entry (e.g. forward packet to a dedicated port, or block packet etc.). By default, the first packet of each flow is sent to the controller, for example, a router simply routes the packet to the controller. After decoding the packet header (examining information like MAC address, or IP address etc.), the controller sets the action field for this particular flow. The controller also informs the router of these actions. Table 1 shows a sample flow table in an OF-Switch. Each flow has packet size N. If there are k flows, the first packet of each flow is redirected to the controller. The controller then decides the action based on header information from the first packet, e.g. all the packets from the first flow are routed to R_2 while the rest of the flow is directed towards R_3.

In this research, we have used OpenFlow-enabled SDN to simulate mobile IP protocols.

Table 2. Address tables of devices

Device name	Address
$Node_1$ (CN)	1.0.4.2
AP_1 (HN)	1.0.2.1
$Node_2$ (HA)	1.0.2.2
MN	1.0.2.3
AP_2 (FN)	1.0.3.1
Node4	1.0.3.2

4   SDN-Enabled Mobile IP

In this section, we discuss implementation of SDN-enabled mobile IP protocols in Estinet simulator [13,14]. We have compared following four protocols: (i) Legacy MIPv6, (ii) PMIPv6, (iii) HMIPv6, and (iv) Network Assisted HMIPv6.

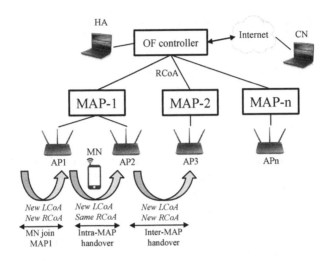

Fig. 5. SDN-enabled HMIPv6 and NA-HMIPv6 implementation

4.1 Simulation Environment

The implementation of SDN-enabled MIPv6 is presented in Fig. 4. Access point AP_1 is the PoA for MN residing in its home network. The CN ($Node_1$) is sending data to MN. The data is first sent to the controller, which directs the route CN-to-MN through AP_1. When MN leaves the communication range of AP_1 and enters into the range of AP_2, it receives a temporary CoA. All the forthcoming data from CN will now buffer into HA until MN updates its new CoA information to HA ($Node_2$). After binding-update is completed, HA sends buffered data to AP_1, which directs data through the router to AP_2 to MN. This process is called tunneling, which continues until the session ends. Tunneling stops when MN returns to home network, or when it enters a new FN. Table 2 shows all the IP addresses associated with the communicating nodes in our simulation. The architectural implementation of MIPv6 and PMIPv6 required minor changes. For example, MAG in PMIPv6 is analogous to HA in MIPv6.

We ran simulations for HMIPv6 setup shown in Fig. 5. Recall that the motivation behind using hierarchical MIPv6 is to benefit from the layered architecture to reduce signaling overhead. The hierarchical architecture comprises of three layers: OF controller (OFC), MAPs, and APs. All the APs in a domain are listed inside OFC. As MN moves, OFC configures LCoA and RCoA for MN given the prefix of the new PoA. In our simulation settings, we have considered three kinds of handovers. In the first instance, a MN joins a network (Join Handover). The MN is assigned a new LCoA and RCoA as it joins MAP-1. When MN leaves communication range of AP_1 to enter the range of AP_2 (intra-MAP handover), the MN is assigned a new LCoA, while the RCoA remains the same. The third instance is when MN leaves the domain of MAP-1 to enter the domain of MAP-2 (inter-MAP handover). MN is assigned a new LCoA and RCoA in this case. The binding between RCoA and LCoA is managed by MAP.

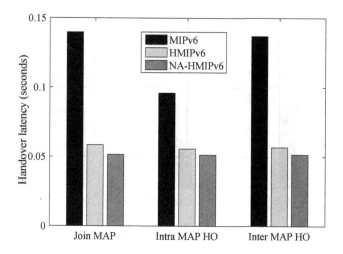

Fig. 6. Comparison of handover latency

Table 3. Handover latency and signaling between MIPv6 and PMIPv6

Protocol	MIPv6	PMIPv6
Total time (sec)	0.1802	0.0068
Number of steps	10	14

Later, we also implemented the Network Assisted HMIPv6 (NA-HMIPv6) in which the MN will initiate the handover process as in HMIPv6. The goal here is to minimize the MN involvement in the handover process while still maintaining the layered architecture of HMIPv6. The architecture for NA-HMIPv6 is same as shown in Fig. 5.

4.2   Simulation Results

Here we examine the performance of SDN-enabled mobility protocols based on the time required to complete the handover, and the number of steps required for this purpose. As mentioned in Sect. 4.1, we have considered three types of handovers in our simulation for HMIPv6 and NA-HMIPv6. Figure 6 shows that SDN-enabled NA-HMIPv6 reduces the handover latency compared to SDN-enabled HMIPv6 and MIPv6.

Since PMIPv6 and HMIPv6 are based on different architecture, direct comparison between two protocols is not possible. Similarly, comparing MIPv6 and PMIPv6 in Table 3, it is also evident that SDN-enabled PMIPv6 performs better than SDN-enabled MIPv6.

Figure 7 shows the number of steps required to complete the handover process. It can be observed that legacy MIPv6 takes lesser number of steps compared to HMIPv6, NA-HMIPv6 and PMIPv6 (see Table 3). However, in our simula-

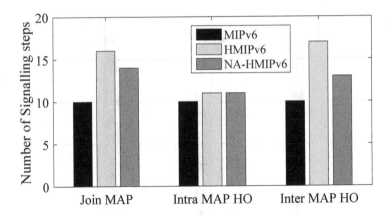

Fig. 7. Comparison of steps required to complete handover

tion, it was observed that the cumulative latency to accomplish handover with MIPv6 was higher than its variants (i.e. PMIPv6, HMIPv6 and NA-HMIPv6).

5   Conclusion

In this paper, we have examined the performance of different mobile IP protocols with software defined network. We have done extensive simulation using Estinet simulator to demonstrate our results. It was found that SDN-enabled PMIPv6 takes lesser time to complete the handover process than MIPv6 protocol. Similarly, handover latency for SDN-enabled MIPv6 is higher when compared to SDN-enabled HMIPv6 and NA-HMIPv6 in different mobility scenarios. Furthermore, NA-HMIPv6 performs better compared to legacy HMIPv6 in terms of handover latency. On the other hand, SDN-enabled MIPv6 takes least number of steps to complete the handover process. A detailed performance comparison between SDN-enabled HMIPv6 and SDN-enabled PMIPv6 needs further investigation and is considered as part of our future research.

Acknowledgement. This work has been funded through Internet New Zealand's research grant of 2016/2017.

References

1. Koodli, R.S., Perkins, C.E.: Mobile Inter-networking with IPv6: Concepts, Principles and Practices. Wiley, Hoboken (2007)
2. Johnson, D., Perkins, C., Arkko, J.: Mobility support in IPv6. Technical report (2004)
3. Guimaraes, C., Corujo, D., Aguiar, R.L., Silva, F., Frosi, P.: Empowering software defined wireless networks through media independent handover management. In: Global Communications Conference (GLOBECOM). IEEE (2013)

4. Farhady, H., Lee, H., Nakao, A.: Software-defined networking: a survey. Comput. Netw. **81**, 79–95 (2015)
5. Jeon, S., Guimarães, C., Aguiar, R.L.: SDN-based mobile networking for cellular operators. In: Proceedings of the 9th ACM Workshop on Mobility in the Evolving Internet Architecture, pp. 13–18. ACM (2014)
6. Wang, Y., Bi, J.: A solution for IP mobility support in software defined networks. In: 2014 23rd International Conference on Computer Communication and Networks (ICCCN), pp. 1–8. IEEE (2014)
7. Hasan, S.F.: A discussion on software-defined handovers in hierarchical MIPv6 networks. In: 2015 IEEE 10th Conference on Industrial Electronics and Applications (ICIEA), pp. 140–144. IEEE (2015)
8. Basu, D., Hussain, A.A., Hasan, S.F.: A distributed mechanism for software-based mobility management. In: 2016 7th IEEE International Conference on Software Engineering and Service Science (2016)
9. Gundavelli, S., Leung, K., Devarapalli, V., Chowdhury, K., Patil, B.: Proxy mobile IPv6. Technical report (2008)
10. Li, Y., Wang, H., Liu, M., Zhang, B., Mao, H.: Software defined networking for distributed mobility management. In: 2013 IEEE Globecom Workshops (GC Wkshps), pp. 885–889. IEEE (2013)
11. Soliman, H., Castelluccia, C., Elmalki, K., Bellier, L.: Hierarchical mobile IPv6 (HMIPv6) mobility management. Technical report (2008)
12. McKeown, N., Anderson, T., Balakrishnan, H., Parulkar, G., Peterson, L., Rexford, J., Shenker, S., Turner, J.: OpenFlow: enabling innovation in campus networks. ACM SIGCOMM Comput. Commun. Rev. **38**(2), 69–74 (2008)
13. EstiNet Technologies Inc.: Estinet Simulator (2017). http://www.estinet.com/ns/
14. Wang, S.-Y., Chou, C.-L., Yang, C.-M.: Estinet openflow network simulator and emulator. IEEE Commun. Mag. **51**(9), 110–117 (2013)

Dynamic Spectrum Management in 5G: Lessons from Technological Breakthroughs in Unlicensed Bands Use

Fernando Beltrán[1](✉), Sayan Kumar Ray[2], and Jairo Gutiérrez[3]

[1] University of Auckland, Auckland, New Zealand
f.beltran@auckland.ac.nz
[2] Manukau Institute of Technology, Auckland, New Zealand
sayan.ray@manukau.ac.nz
[3] Auckland University of Technology, Auckland, New Zealand
jairo.gutierrez@aut.ac.nz

Abstract. This paper discusses a number of issues associated with the increasing need to improve the utilization of unlicensed spectrum as a number of new technological advances provide an opportunity to share scarce resources in a dynamic fashion in the future 5G networks. The growth in connected devices via cellular and Wi-Fi networks is being complemented with a significant increase in networked "things" and this proliferation of devices presents a challenge to Spectrum Authorities. We propose that the ultimate purpose of Dynamic Spectrum Management (DSM) is to improve spectrum usage efficiency by fully exploiting spectrum sharing while assuring minimum undesired interference. Our aim is the identification of economic issues that impact the development of efficient markets for 5G networks that rely on dynamic spectrum technologies in the unlicensed spectrum. The paper covers how technological breakthroughs in spectrum access technologies challenge our current understanding of spectrum management. In each case, the contribution of the paper includes policy proposals or more focused regulatory instruments while the concluding section sums up the paper's key message about the interplay between technology and policy that helps lay out elements that regulators and policy-makers need to attend to when adopting practices that implement Dynamic Spectrum Management in the unlicensed spectrum.

Keywords: Dynamic Spectrum Management
Spectrum allocation and assignment · 5G networks · Unlicensed spectrum
Wi-Fi · LTE · IoT · TVWS · mmWave

1 Introduction

This discussion paper captures the most relevant aspects that need to be considered for regulation and policy to deliver socially optimal outcomes in a wireless service environment characterised by the dynamic use of the radio spectrum. As the utilisation of the radio spectrum evolves from exclusive band operation to shared, opportunistic, and intermittent usage, such new spectrum access modes demand a dynamic approach to

© ICST Institute for Computer Sciences, Social Informatics and Telecommunications Engineering 2018
P. H. J. Chong et al. (Eds.): SmartGIFT 2018, LNICST 245, pp. 250–259, 2018.
https://doi.org/10.1007/978-3-319-94965-9_25

access and interference control and so must spectrum policy and management, originally conceived as a command-and-control regime. Our aim is the identification of economic issues that impact the development of efficient markets for 5G networks that rely on dynamic spectrum technologies in the unlicensed spectrum.

After discussing the supporting policy principles of spectrum management in Sect. 2, the paper steers the discussion toward how technological breakthroughs in spectrum access technologies in the unlicensed spectrum challenge our current understanding of spectrum management. In particular, in Sect. 3 we analyse the impact that Long Term Evolution (LTE) may bring in as it is proposed to be deployed on Wi-Fi bands. Section 4 follows by discussing Internet of Things (IoT)'s need for new spectrum; then in Sect. 5 advances in opportunistic access and utilisation of TV White Spaces (TVWS) are presented. In Sect. 6 we discuss the irruption of Millimeter Wave technology. In each case, the contribution of the paper includes policy proposals or more focused regulatory instruments, all of which are intended to align with the issues the paper raises. Our concluding section sums up the paper's key message about the interplay between technology and policy that helps lay out elements that regulators and policy-makers need to attend to when adopting practices that implement Dynamic Spectrum Management in unlicensed spectrum for future 5G networks.

2 Spectrum Management

As radio spectrum is assigned as usage rights over ranges of frequencies called bands, undesired spill-over signals from using the spectrum in one band, known as interference, may negatively impact its usage in adjacent bands. Hence, the main reasons for spectrum management: band allocation and interference minimization. Rights to transmit are usually allocated to users in the form of a license, terms and conditions of which should not lead to excessive interference.

In light of technology and policy advancements Cave and Webb [1] discuss new features of spectrum management which, in addition to the conventional approach, the authors argue, needs to provide assurance that the value of spectrum to society is maximized. The latter calls for making a key objective of spectrum management that it allows as many users to gain efficient access to the spectrum as possible. Therefore, spectrum needs to be managed because, with current technology, maximizing its value to society can only be achieved by coordinating who can use which bands and over which geographical extensions.

A Spectrum Authority (SA) performs its spectrum management mandate by, first assessing all potential uses of a band and deciding about the type of use the band will be dedicated to; this is known as **spectrum allocation**. Then the SA provides a license to one or more operators for exploitation of the radio band; this is known as **spectrum assignment**.

Spectrum allocation and spectrum assignment are centrepieces of spectrum management. Spectrum users need assurance about the conditions under which spectrum is used; such conditions are paramount to achieving technical efficiency and minimization of interference. Long-term licenses provide the stability users seek and allow a SA to achieve other goals: conditions on service availability and coverage requirements imposed on the licensees. In a few cases SAs have decided specific bands do not require licensing but only the adherence to some basic technical requirements, typically enforced to minimize interference. The utilization of this type of unlicensed bands may be rather uncoordinated.

Technological innovation in wireless transmission is proving the static features of spectrum management need to be thought over. Policy decisions that respond to increasing pressure for competition in leading communication sectors such as cellular telephony and wireless broadband services are also causing spectrum management to evolve. When static features of spectrum management are changed or disrupted by technological innovations and policy changes, it is clear that managing the spectrum needs to cater for new challenges.

Technology changes and market interests exert pressure on SAs, demanding more spectrum and a more diverse approach to allocation and assignment. Leading SAs across the world have started a shift towards the inclusion of new elements and tools that promise to provide spectrum management with the ability to respond more dynamically to the changes technology brings in and users demand.

One of the techniques that spite being already allowed for quite a long time is retaking a central stage in spectrum management considerations is Spectrum Sharing (SS). A spectrum utilization scheme that allows two or more parties to utilize the same range of frequencies while no exclusivity is granted to any of them, SS is also a renewed tool for spectrum management. Techniques that facilitate spectrum sharing are divided into uncoordinated – radio systems adjust their operation to coexist with other radio systems with little information to share, and coordinated – techniques that require coexisting radio-frequency (RF) systems to exchange information to share the same frequency band. Examples of the former are dynamic channel selection and adaptive frequency hop to listen-before-talk, whereas examples of the latter include multiplexing techniques such as FDMA, TDMA or CDMA or channel-based control methods, such as CSMA/CA.

Thus, a renewed approach to spectrum management must acknowledge the most important changes in technology and policy, the economic importance of spectrum (value), changes in its utilization (innovation), and the need for efficient utilization in a market-driven way, all of which must be considered while managing the social role of spectrum. Allowing spectrum-sharing arrangements challenges the conventional regulatory approach to commercial use of the spectrum in particular for mobile telecommunications services.

Although spectrum sharing is favoured by many observers and seems to be finding a clear way as a policy tool of SAs, not all bands can or should be potential sources of sharing. The SCF Report [3] indicates that in Europe bands for distress calls, maritime navigation, and air traffic control must remain exclusive, deeming these bands as non-shareable. In spite those pockets, SCF concludes that "it is possible in many areas of the spectrum currently under commercial or administrative licensing regimes to use a shared regime without endangering those other services vital to safety of life".

The natural progression of spectrum management that aims to account for the features described and discussed above is called Dynamic Spectrum Management, DSM. The ultimate purpose of DSM is to improve spectrum usage efficiency by fully exploiting spectrum sharing while assuring minimum undesired interference. A number of current issues that demand spectrum management to be modified, changed or adapted for future 5G networks will next be discussed. In all cases the characteristics of those technology breakthroughs challenge aspects of the conventional approach to spectrum management, demanding a decidedly DSM-oriented context.

3 LTE vs WiFi in Unlicensed Spectrum

Telecommunications vendors and the cellular networks operators championing LTE are targeting the unlicensed 5 GHz bands, currently being used by Wi-Fi, Zigbee and few other communication systems, to expand the LTE capacity and meet traffic growth. Here we briefly discuss some relevant aspects of LTE's history in the unlicensed spectrum. In 2013 Qualcomm proposed the idea of deploying LTE in the unlicensed spectrum [4]. Two versions, LTE-U, which is the pre-standard proprietary version backed by the LTE-U Forum, and Licensed Assisted Access (LAA), which was developed by 3GPP, are the contenders.

Initial deployment of these LTE versions in the unlicensed bands is expected through small cells for DL only and then slowly for UL as well. Both LTE-U and LAA utilize carrier aggregation functionality using both the unlicensed bands and the licensed spectrum. While LAA complies with the different regulatory requirements for the usage of the unlicensed spectrum, LTE-U does not and instead uses the duty cycling-based system called Carrier Sensing Adaptive Transmission (CSAT). An LTE-U cell using CSAT does not sense the occupancy of a channel before transmitting and instead turns its signal on and off over small periods of time to, respectively, occupy the channel to transmit and vacate the channel for other technologies like Wi-Fi. LTE-U's focused deployment options are only in the non-LBT (Listen-Before-Talk) required regions in the world [5, 6].

The LTE-U Forum is progressing with developing the protocols for LTE-U operations in the 3.5 GHz band, while T-Mobile (US) is looking to adopt LAA for that same band. However, efficient operation in the 3.5 GHz band will set a strong requirement for low power RF equipment, e.g., low power small cell technologies for both LTE-U and LAA.

Additionally, improvements to the LTE standards, in Release 14, include enhanced LAA (eLAA), which among other functions provides full support for UL transmissions in the unlicensed spectrum. The issue becomes the potential overutilization of offloading and hence its impact on many Wi-Fi services. LTE-based networks can easily switch from unlicensed to licensed use, an ability that is not available to Wi-Fi networks and its many users.

The advances discussed above point towards mechanisms to enable the coexistence of LTE and Wi-Fi networks with convergence on both scheduled and ad-hoc wireless configurations. Regardless of the techniques available, the SAs need to address the issue of whether the co-existence scenario delivers more value than the existing unlicensed Wi-Fi scenario or not.

4   Internet of Things

The Internet of Things (IoT) refers to the widespread use of systems, heterogeneous technologies and the evolving paradigm of the interconnectedness of devices, using TCP/IP protocols, around our physical environments. IoT includes a new wave of sensor devices and interoperates with the growing cloud network infrastructure. On the long run it is envisioned that an IoT ecosystem will facilitate the interaction of devices (mobile or fixed), smart objects and other real world devices just as humans interact nowadays using internet-based applications. The IEEE is currently working on a reference architecture, which will define the basic IoT architectural building blocks and how they could be seamlessly combined into multi-tiered systems [7].

Currently spectrum allocations tend to be IoT application-specific and must satisfy the service requirements of individual applications. IoT allocations are particularly active in the Sub-6 GHz spectrum. Generally speaking, service requirements range from excellent and ubiquitous coverage, ultra-low power operations, provision of adequate bandwidth, to secured and low cost communication and guaranteed message delivery. While, from a spectrum allocation perspective, it is a challenge to meet these varied requirements, an initial step is making available globally harmonized low-frequency spectrum in the unlicensed bands, e.g., bands below 1 GHz like 870–876 MHz and 915–921 MHz along with the TV white spaces [8, 9]. In future, 700 MHz bands may also become available. All these bands allow extended coverage and support interconnection of a higher number of less complex and low-powered IoT devices. These are also beneficial to run IoT applications, which require in-building penetration. SigFox, LoRa (Long Range) and NB-IoT (Narrowband IoT) are examples of notable narrowband Low-power WAN (LPWAN) technologies operating in unlicensed spectrum. Low frequency bands are, however, scarce and high in demand, so there should be ways to find and free more such globally harmonized bands that can be made available for IoT applications. Another notable example is the Wi-SUN (Wireless Smart Ubiquitous Network) technology, which is based on the IEEE 802.15.4g standard. Traditionally SA's allocation and assignment processes have favoured exclusivity. With an inability to foresee the pathways of technological innovation a SA needs to reassess the importance of modifying the assignment stage to favour unlicensed or shared bands and hence alleviate scarcity.

Apart from these, some interests on shared bands over 2 GHz are also there, particularly for applications with higher bandwidth requirements, like video monitoring. Such bands include 2.3 GHz, 2.4 GHz, 2.6 GHz, 3.4–3.8 GHz, and 5 GHz. However, with multiple wide-ranged IoT technologies flocking the unlicensed spectrum, interference may always be an issue with the increase in the number of IoT devices even if the devices are low-powered. Recent research indicates that a license-exempt model, in fact a way of skipping the assignment phase in spectrum management, facilitates the rapid development of IoT devices as it eliminates the need for time-consuming negotiations about the spectra to be used. This could directly result in cheaper IoT nodes [10]. Another possibility could be setting a worldwide default frequency in the range of 915–928 MHz for IoT devices to facilitate compliancy and deployment. Other opinions have voiced the requirements for making the IoT devices themselves understand which country they are operating in and what are the available spectrum bands there and operate accordingly [11].

5 TV White Spaces

TVWS refers to frequencies allocated to licensed Digital Terrestrial TV (DTT) broadcasting services that are unused and freed up for unlicensed radio devices known as White Space Devices (WSD). These devices can dynamically share and opportunistically use TVWS on a secondary basis without interfering with each other or with the primary licensed service providers (digital TV broadcasters and wireless microphone users). TVWS, an important cognitive radio application, enables long-range services in broadband speeds. The first major implementation of the concept of Dynamic Spectrum Access (DSA) has been in the TVWS spectrum bands [12]. DTT broadcasting uses the VHF band (30 MHz to 300 MHz) and lower part of UHF spectrum bands (300 MHz to 1000 MHz) [13]. Table 1 lists some of the countries that have TVWS regulations in place [12, 13]. The TVWS spectrum market will experience the coexistence of WSDs and services of the different unlicensed technologies, including the IEEE 802.22 Wireless Regional Area Network (WRAN), IEEE 802.11af, IEEE 802.15.4m, ECMA-392 and Weightless, through dynamic sharing of the available spectrum bands. Trials and deployments of these are underway in multiple countries.

Table 1. Some of the countries with TVWS usage regulations.

Countries	Usage bands	Regulatory body
USA	VHF: 54–88 MHz and 174–216 MHz UHF: 470–698 MHz	Federal Communications Commission (FCC)
UK	UHF: 470–790 MHz	Ofcom
Singapore	VHF: 174–230 MHz UHF: 470–806 MHz	Info Communications Development Authority (IDA)
Canada	UHF: 470–698 MHz	Industry Canada
Europe	UHF: 470–790 MHz	European Communications Commission

Primarily, TVWS usage aims to enable secondary users using the bands without interfering with the primary incumbent users. Unlicensed Shared Access of spectrum is the possible approach to follow for DSM in this case. A strong need is there to protect the existing investments and users in the TV bands and calls for a globally coordinated and holistic approach to deal with key issues, like, identifying TVWS spectrum in different regions and countries, non-harmonized specifications for WSDs and lack of global standards or regulatory frameworks for TVWS usage. Geolocation databases are globally accepted as the most promising solution to identify and use TVWS spectrum for a variety of services. They store information regarding operating frequencies, schedules and locations of the licensed DTT providers and other users and devices sharing the TV bands. The WSDs can access the list of currently available TVWS channels in a region by providing their own geolocations to the geolocation databases in that region. Not having globally harmonized TVWS regulation for geolocation databases and WSDs, however, pose few challenges. It will not be possible to readily use an auto-configurable WSD from one region to work in another. For example, a WSD from UK (supporting only UHF TVWS bands) will not readily work everywhere in US, which supports both VHF and UHF bands. In addition, types of supported WSDs vary between countries. While US supports sensing-only WSDs, other countries do not support them. Moreover, to introduce new TVWS technologies or services that will not interfere with primary services or other coexisting services in a region, the databases used need to have common technical standards to identify and accommodate the new TVWS technologies or services and related WSDs introduced. Thus, for efficient DSM of TVWS bands in unlicensed bands, there needs to be standardised policies and regulations enabling the harmonization of geolocation databases and use of WSDs worldwide. Although there exists the European harmonized standard (ETSI Harmonized Standard) for WSDs, it is only a voluntary scheme [13].

6 Millimeter Wave (mmWave)

The millimetre wave (mmWave) refers to frequency spectrum above the 24 GHz bands that may range up to 300 GHz. It can cater for high broadband capacity and is emerging as one of the promising technologies for 5G communication offering a large pool of available spectrum for mobile users, satellite users, and other commercial users to share and coexist. The recent FCC mandate has opened up nearly 11 GHz of high frequency spectrum in the mmWave bands for fixed and mobile broadband usage of which 3.85 GHz is licensed and 7 GHz is unlicensed spectrum [14]. These 7 GHz of unlicensed bands combined with the already existing 57–64 GHz of unlicensed spectrum, will provide 14 GHz of contiguous spectrum for unlicensed usage in the mmWave bands, which will be nearly 15 times more than the WiFi unlicensed spectrum in lower bands. Moreover, in the US, there will be 600 MHz of spectrum for dynamic shared access in the 37–37.6 GHz bands for commercial and federal users [14]. The UK has made available 18.3 GHz of unlicensed spectrum in the 60–80 GHz bands, while in Europe the 57–64 GHz band is for licensed-exempt usage.

Harmonization of the mmWave bands worldwide in regards to usage models is crucial as opportunities exist for coexistence of different types of technologies sharing the bands. Sharing of the unlicensed spectrum in the mmWave bands should be done in a way to allow for substantial spectrum reuse while still keeping the interference to the lowest. Recent research on spectrum sharing in the mmWave bands has reported considerable performance enhancement utilizing concepts like uncoordinated spectrum sharing and hybrid spectrum access [15]. However, a consensus has neither been reached as of yet on a globally accepted coexistence model in the mmWave bands nor been reached on global usage policies in these bands. In addition, regulatory frameworks in the mmWave bands vary between regions and this may hinder spectrum sharing. For example, technical implementation specifications for antenna arrays can hugely vary between bands that are gigahertz apart and this will require user devices to not only just operate in multiple bands but also to self-identify the underlying operational band and to customize its configuration accordingly. This would require a global harmony or standardization of the technical specifications of the equipment and devices operating in the unlicensed parts of the mmWave bands.

Efficient utilization of available unlicensed spectrum in mmWave bands can enable the coexistence of heterogeneous deployments. Dynamic spectrum sharing will be important in such scenarios. It may be possible to adjust the amount of utilized spectrum in real-time depending on the service demand at any given moment, while still considering the geographical characteristics [14]. Thus, to enable coexistence of different technologies and services in the unlicensed mmWave bands, dynamic spectrum sharing is important and global harmony needs to be reached in terms of usage models, technical specifications of equipment, operational regulations, dynamic spectrum sharing mechanisms, and allocating priority access to services/operators depending on demand etc., before mmWave in the unlicensed bands can be successfully commercialized for future 5G networks.

7 Moving Towards DSM

The above sections discussed the different technological breakthroughs in spectrum access technologies in the unlicensed spectrum. Quite a number of issues associated with the increasing need to improve the utilization of the unlicensed spectrum are highlighted for regulators and policy-makers to consider when adopting practices to implement Dynamic Spectrum Management for future 5G networks.

As LTE-U and LAA seek to be deployed on the 5 GHz band, either as a mixed U/L implementation or fully unlicensed, such as eLAA, the question of fairness arises as the most critical coexistence issue for SAs. With 5 GHz not being a greenfield spectrum, its management needs to answer what policies are there to govern such fair sharing of spectrum. Clearly the initial allocation of the band is now being questioned, somehow, as new technology and commercial interests are pushing its way into the band. This situation exemplifies the threat that unlicensed bands face as the assignment problem was never really solved in terms of the definition of property rights; after all, 5 GHz as well as 2.4 GHz are conspicuous examples of spectrum commons, which foresees the high possibility of overuse in the unlicensed bands. Although preliminary research results have shown that Wi-Fi performance is not degraded in the presence of LTE-U and LAA, there are still some concerns about the potential dominance of LTE and Wi-Fi in the unlicensed bands.

One of the main concerns about the expansion of IoT technologies is how to effectively achieve worldwide default (preferably) unlicensed spectrum allocation for their operation. A rising challenge for SAs is the identification of internationally accepted mechanisms that allow IoT devices to understand which country they are operating in and self-switch to allocated (legal) IoT spectrum bands. The most likely scenario for a world of IoT is one in which multiple technologies use unlicensed spectrum in an uncoordinated manner. Assuming that the rise of universally accepted

LPWAN standard will include evolving into a support for M2M/IoT connections worldwide, any SA would need to be concerned with how interference between different wide-ranged LPWAN technologies will be handled. Foreseeing the complexity of yet another spectrum commons scenario, which might include default IoT frequencies allocated worldwide to facilitate global roaming and seamless connectivity on, for instance, the 915–928 MHz band, 'fair sharing' of such unlicensed spectrum among IoT technologies erects itself as a critical issue.

TVWS is intended to mitigate spectrum-sharing challenges. International experience indicates different countries maintain different compliancy regulations to assess the suitability and functioning of geolocation databases, which are regulated by national SAs but mostly provided and maintained by private providers. In such cases, DSA for TVWS bands would raise concerns that mostly deal with the provider, such as its suitability, the criteria it should fulfill, and the verification of its fair access and opportunity policies to use white bands on a temporary basis by opportunistic, second-tier operators. Also, when bands of choice in a region are not available to WSDs, band diversity may be a potential solution where the network and its WSDs are able to operate across multiple bands so that there are always bands available. This requires the geolocation database operators across regions to update their respective databases periodically with appropriate information and to share the updates amongst each other.

Additionally, the challenges associated with mmWave are characterized by the uncertainties linked to this newer technology. There are areas in common with the technologies discussed above: the global harmonization of bands, the search for an agreement on a coexistence model for the future use of these unlicensed bands by different technologies, and the appropriate role for the SAs. mmWave has some other concerns that are associated to its incipient nature, i.e., technical specifications of equipment, operational regulations, dynamic spectrum sharing mechanisms, and allocating priority access to services in the unlicensed spectrum depending on demand.

Finally, a preliminary proposal is to use some of the principles that have been successful in the provision of network management facilities using software defined networking [16]. As mentioned before one of the key areas in common is the search for an agreement on the use of the unlicensed bands (UB) by the different technologies surveyed. We propose a **Spectrum Controller** mechanism, akin to the Software Controller in software-defined networks (SDNs), which will coordinate UB requests within a given autonomous system (AS). Following the principles of SDNs, the requests are made via a control plane (or channel) and the associated allocation takes place in the assigned data channel. Upon termination of the connection, the SC will update the register of shared resources available for future requests. The SC is therefore the entity, which keeps track of the resources shared within an AS by a number of significantly different technologies which have in common a need to dynamically shared unlicensed spectrum. An area of future research is to work on the coordination of spectrum usage among cooperating autonomous systems.

References

1. Cave, M., Webb, R.: Spectrum Management-Using the Airwaves for Maximum Social and Economic Benefit, 1st edn. Cambridge University Press, Cambridge (2016)
2. Hossain, E., Niyato, D., Han, Z.: Dynamic Spectrum Access and Management in Cognitive Radio Networks, 1st edn. Cambridge University Press, Cambridge (2009)
3. Forge, S., Horvitz, R., Blackman, C.: Perspectives on the value of shared spectrum access-final report for the european commission, Technical report (2012)
4. LTE in Unlicensed Spectrum: Harmonious Coexistence with Wi-Fi. White Paper, Qualcomm Research (2014)
5. Mobile Broadband Evolution towards 5G: Rel 12 & Rel-13 and Beyond. White Paper, 4G Americas (2015)
6. Beltran, F., Ray, S.K., Gutierrez, J.: Understanding the current operation and future roles of wireless networks: co-existence, competition and co-operation in the unlicensed spectrum bands. IEEE J. Sel. Areas Commun. **34**(11), 2829–2837 (2016)
7. P2413 - Standard for an Architectural Framework for the Internet of Things (IoT). IEEE Standards Association (2014)
8. The Future Role of Spectrum Sharing for Mobile and Wireless Data Services: Licensed Sharing, WiFi and Dynamic Spectrum Access. Statement, Ofcom (2014). https://www.ofcom.org.uk/consultations-and-statements/category-1/spectrum-sharing. Accessed 25 Mar 2018
9. Draft BEREC Report on Enabling the Internet of Things. BEREC report. 141 (15), BoR (2015)
10. Regulation and the Internet of Things. GSR Discussion paper (2015)
11. ACMA Spruiks Default IoT Spectrum Worldwide. http://www.zdnet.com/article/acma-spruiks-default-iot-spectrum-worldwide. Accessed 25 Mar 2018
12. Webb, W.: Dynamic White Space Spectrum Access. Kindle edn. Webb Search Limited (2013)
13. Implementing TV White Spaces. Statement, Ofcom (2015)
14. 47 CFR Parts 2, 25, 30, et al.: Use of Spectrum Bands above 24 GHz for Mobile Radio Services; Proposed Rule. Report, Part IV. 81 (164). Federal Communications Commission (2016)
15. Gupta, A.K., Andrews, J.G., Heath Jr., R.W.: Can operators simply share millimeter wave spectrum licenses?. In: Information Theory and Applications (ITA) Workshop, San Diego. IEEE (2016)
16. Modieginyane, K.M., Letswamotse, B.B., Malekian, R., Abu-Mahfouz, A.M.: Software defined wireless sensor networks application opportunities for efficient network management: a survey. Elsevier Comput. Electr. Eng. J. **66**, 274–287 (2018)

Dual Sensing Scheduling Algorithm for Wireless Sensor Network Based Road Segment Surveillance

Farhan Khan[(⊠)] and Sing Kiong Nguang

Department of Electrical and Computer Engineering, The University of
Auckland, Auckland, New Zealand
fkhal13@aucklanduni.ac.nz, sk.nguang@auckland.ac.nz

Abstract. In this paper, a dual sensing scheduling algorithm is proposed which is a modified version of VISA technique for sensing scheduling in road networks where targets can enter from both sides of the road. VISA and similar algorithms are based on the idea of designated entrance points and protection points and are very suitable for military scenarios. In comparison, civilian applications mostly use two-way roads and dual carriageways with entrance points on both ends of the roads calling for a modification of the VISA technique to make it suitable for two-way detection. The proposed algorithm achieves detection on a two-way road by using two parallel scan waves originating from the midpoint sensor on the road segment but in opposite directions. The proposed modification of the VISA algorithm improves the detection time by reducing it to half as compared to VISA but at the cost of decreased network lifetime. The proposed algorithm is also compared to Duty Cycling and Always-Awake schemes.

Keywords: Sensing scheduling · Virtual Scanning · Duty Cycling

1 Introduction

Wireless sensor network based road segment surveillance is one of the key operations in military applications and with the rise of road traffic monitoring applications in civilian contexts, there is a renewed interest in this application area [1].

Traditional methods have mostly focused on full coverage, Always-Awake based techniques [2–5]. Always-awake techniques generally have very limited network lifetime because the sensors do not sleep during network operation, but they provide the smallest average detection time usually denoted by zero in literature. In order to increase network lifetime, the Duty Cycling based approaches allow all sensor nodes to start their sensing operation simultaneously for w seconds and after that the whole network goes to sleep for T seconds [6–9].

One of the earliest and state-of the-art work which utilized scan waves for detection of intruding vehicles was proposed by Jeong et al. and is known as *Virtual Scanning Algorithm* (VISA) [15, 16]. The work proposed in this paper is based on a modification of VISA. VISA uses the concepts of entrance points and protection points in a road

© ICST Institute for Computer Sciences, Social Informatics and Telecommunications Engineering 2018
P. H. J. Chong et al. (Eds.): SmartGIFT 2018, LNICST 245, pp. 260–267, 2018.
https://doi.org/10.1007/978-3-319-94965-9_26

network and sends waves of scan to detect vehicles as shown in Fig. 1. In Fig. 1, the left end of the road segment is the entrance point where the vehicle can get into the road network and the right end of the road segment is protection point which needs to be protected from intruding vehicles. In VISA design, all sensors are waked up one by one for a certain working time w from the direction of protection point towards entrance point after a network-wide silent time. This wave of sensing activities guarantees the detection of target and allows additional sleeping time for individual sensors. Jeong et al. also argue that the virtual scan of the opposite direction i.e., from the entrance point to the protection point cannot guarantee target detection if a very fast target enters just after the start of the network-wide silent time. VISA is very suitable and appropriate for military applications where there is a concept of designated entrance points and des-ignated protection points, but the design of VISA does not consider detection in the case of two-way roads which is the most common type of roads in civilian applications.

This paper proposes a dual sensing scheduling algorithm. The basic idea of dual sensing scheduling algorithm is to ensure detection of vehicles on two-way roads by using two scan-waves in opposite directions initiating from the midpoint sensor of a road segment. In simplest terms, dual sensing scheduling can be thought of as two parallel VISA scan waves but in opposite directions with midpoint sensor as the starting point of both of the scans or in terms of VISA terminology, midpoint sensor can be thought of as a protection point for both of the scans. This is illustrated in Figs. 2 and 3 for the same road segment of length d which is shown in two sub-segments of length $d/2$ each. This type of parallel dual sensing puts quite a stringent demand on network lifetime but greatly reduces the average detection time.

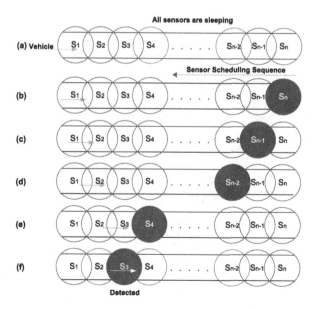

Fig. 1. Basic operation of VISA illustrated for a road segment of length d with entrance point on the left and protection point on the right [15, 16].

The rest of the paper is organized as follows: Sect. 2 briefly discusses related work followed by problem formulation in Sect. 3. In Sect. 4, comparison of analytical network lifetime and average detection time is described, and the paper is concluded in Sect. 5.

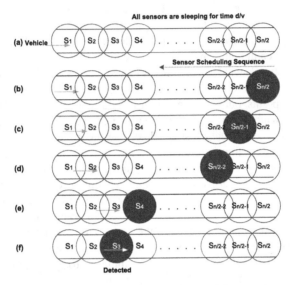

Fig. 2. Basic operation of Dual Sensing Scheduling illustrated for a road segment of length $d/2$ and the midpoint at sensor $S_{n/2}$ for vehicles entering from left side of the road.

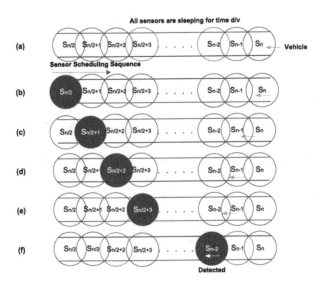

Fig. 3. Basic operation of Dual Sensing Scheduling illustrated for a road segment of length $d/2$ and the midpoint at sensor $S_{n/2}$ for vehicles entering from right side of the road.

2 Related Work

WSN based surveillance algorithms for infrastructure monitoring have mostly paid attention to full-coverage in two-dimensional open spaces [2–5]. In [3], Cardei et al. proposed a method to extend the sensor network life time by organizing the sensors into a maximal number of set covers that are activated successively. Only the sensors from the current active set are responsible for monitoring all targets and for transmitting the collected data, while all other nodes are in a low-energy sleep mode. They modeled the solution as the maximum set covers problem and designed two heuristics that efficiently compute the sets, using linear programming and greedy approach to monitor a set of static targets at known locations. In [10], Rabbat and Nowak presented an approach to source localization and tracking using received signal strength measurements. Based on incremental gradient descent-like optimization methods, their algorithm required small amounts of data to be communicated over short distances. In [11], Chen et al. proposed a fully decentralized, light-weight, dynamic clustering algorithm for target tracking. Instead of assuming the same role for all the sensors, they envisioned a hierarchical sensor network that is composed of two main items: a static backbone of sparsely placed high-capability sensors which will assume the role of a cluster head upon triggered by certain signal events; and low-end sensors whose function is to provide sensor information to cluster heads upon request. A cluster is formed, and a cluster head becomes active, when the acoustic signal strength detected by the cluster head exceeds a pre-determined threshold. The active cluster head then broadcasts an information solicitation packet, asking sensors in its vicinity to join the cluster and provide their sensing information. In [12], Yao et al. presented a localization algorithm based on the observation that signals from different nodes arrive at the target at different times. In [13], Gui et al. proposed a patrolling surveillance algorithm that allows a virtual patrol to move along a predefined path waking up sensors adjacent to the patrol's path according to a schedule in order to track the target. In [14], He et al. proposed a target detection system that allows a group of cooperating sensor devices to detect and track the positions of moving vehicles in an energy-efficient and stealthy manner. They traded off energy-awareness and surveillance performance by adaptively adjusting the sensitivity of the system. In [15, 16], Jeong et al. noted that as compared to two-dimensional open space surveillance, the road network surveillance is different due to two reasons: the first is that the movement of target vehicles is confined within road segments, the second is that the road network maps are normally known in advance. They proposed Virtual Scanning Algorithm (VISA) which has already been discussed in the previous section. They also proposed a hole-handling mechanism within the same work. In [1], Chen et al. proposed a twice deployment node balance (TDNB) algorithm which provides better performance than VISA in terms of network lifetime by dividing the deployment of the sensor nodes into two phases instead of deploying all the sensor nodes at one time.

3　Problem Formulation

The problem is to ensure that all intruding targets from both sides of the road on a two-way road segment are detected before they reach midpoint of the road segment keeping in mind to achieve a suitable lifetime for the sensor network. The next section describes the dual sensing scheduling algorithm along with performance comparison with other techniques.

4　Dual Sensing Scheduling Algorithm for Road Segment Surveillance

The assumption is that n sensors are placed on a road segment of length d. Each sensor has a sensing radius of r which is sufficient to scan width of the road. Let w be the minimum working time needed by a sensor in order that the sensor can reliably detect a target. Let v be the maximum target speed. The targets can enter from both sides of the road segment. The traditional full coverage algorithms where sensors remain turned on all the time are called *Always-Awake*. A better design can be built based on the observation that it takes at least d/v seconds for a target to pass a road segment of length d at a maximum speed v. Therefore, all sensors in the road segment can sleep together for d/v seconds, which is defined as the silent time of the road network. After this silent time, all nodes wake up simultaneously for detection. This technique is called *Duty Cycling*. The *VISA* technique is shown in Fig. 1. After all sensors sleep for d/v seconds, sensors are waked up one by one for working time w from the rightmost sensor S_n toward the leftmost sensor S_1. This scan-wave of sensing activities guarantees the detection [15, 16] as shown in Fig. 1.

The dual sensing scheduling algorithm as shown in Figs. 2 and 3 initiates two parallel scan-waves from the midpoint of the road segment d approximately designated by sensor position $S_{n/2}$. The objective of dual sensing algorithm is to ensure detection of vehicles entering from both sides of the road segment. This feature of dual sensing algorithm provides detection of target vehicle in half of the time as compared to VISA. This is discussed in next two subsections.

4.1　Analytical Network Lifetime Comparison

In order to compare the sensor network lifetime, the parameters used by Jeong et al. in [15, 16] are being used as it is. These parameters are shown in Table 1. In Table 2, overall analytical results for all the four techniques are presented.

Figure 4 shows the comparison of lifetime among the four techniques. The road segment considered in this figure is a 1000 m in length with 300 sensors deployed on the road side for detecting vehicles coming from both direction. The assumed vehicle speed is 64 km/h (40 miles/h). For example, for $w = 5$ s, VISA has a lifetime of 43.57 h, Dual Sensing 11.28 h, Duty Cycling 1.71 h and Always-Awake 0.14 h.

Table 1. Parameters for analysis [15, 16].

Parameter	Definition
T_{life}	*Lifetime that a sensor can work continuously corresponding to its energy budget*
T_{net}	*Sensor network lifetime*
T_{work}	*Working time that a sensor needs to work for reliable detection. Normally $T_{work} = w$*
T_{sleep}	*Sleeping time of each sensor*
T_{scan}	*Scan time that a virtual scan wave moves along the road segment. $T_{scan} = nw$*
T_{silent}	*Silent time that the whole sensor network remains silent; that is, time that a target passes through the road segment of length d. $T_{silent} = d/v$*
T_{period}	*Schedule period of the sensor network. $T_{period} = T_{scan} + T_{silent}$*

Table 2. Performance analysis of four techniques.

Technique	Sleeping (T_{sleep})	Working (T_{work})	Network lifetime (T_{net})	Avg. detection time
Always-Awake	0	T_{life}	T_{life}	0
Duty Cycling	d/v	w	$(T_{life}/w)(w + d/v)$	$d^2/(2v(wv + d))$
VISA	$(n - 1)w + d/v$	w	$(T_{life}/w)(nw + d/v)$	$d/2v$
Dual Sensing	$((n/2) - 1)w + d/v$	w	$(T_{life}/2w)((n/2)w + d/v)$	$d/4v$

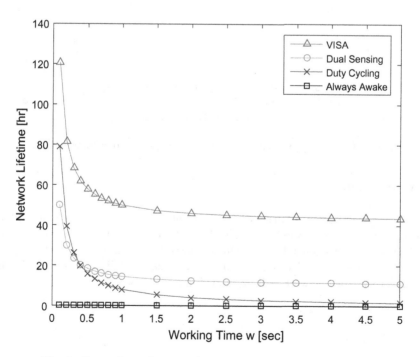

Fig. 4. Comparison of network lifetime according to working time *w*.

4.2   Analytical Detection Time Comparison

Figure 5 compares the average detection time after a target enters a road segment among the four techniques. VISA detects with a constant delay $d/2v$ as proved in Appendix A of [16] and Dual Sensing with a constant delay of $d/4v$ regardless of the working time w. For example, for $w = 5$ s, VISA detects target within 28.12 s, Dual Sensing does within 14.06 s, Duty Cycling does within 25.82 s and Always-Awake does without any delay. Therefore, Dual Sensing outperforms Duty Cycling as well as VISA in terms of average detection time.

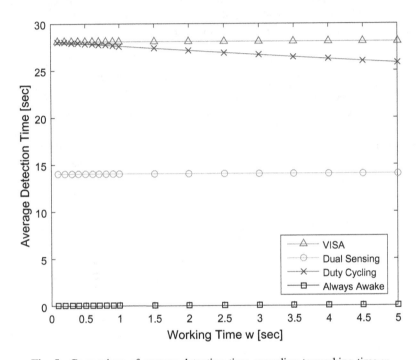

Fig. 5. Comparison of average detection time according to working time w.

5   Conclusion

This work has presented a Dual Sensing Scheduling algorithm which is a modified version of state-of-the-art VISA technique. Using the concepts mainly derived from the VISA technique, the distinct feature of the proposed algorithm is that it provides target detection for two-way road segments with constant delay. Application of Dual Sensing on a larger size road network will be studied in a future work.

References

1. Chen, L., Li, Y., Li, Z., Li, W., Su, P., Hua, C., Luo, Q., Yin, F., Jiang, Y.: Twice deployment node balance algorithm for road network surveillance. Int. J. Distrib. Sens. Netw. **10**(5), 957851 (2014)
2. Abrams, Z., Goel, A., Plotkin, S.: Set K-cover algorithms for energy efficient monitoring in wireless sensor networks. In: Proceedings of the IEEE International Conference on Information Processing in Sensor Networks (2004)
3. Cardei, M., Thai, M.T., Li, Y., Wu, W.: Energy-efficient target coverage in wireless sensor networks. In: Proceedings of IEEE INFOCOM (2005)
4. Kumar, S., Lai, T.H., Balogh, J.: On K-coverage in a mostly sleeping sensor network. In: IEEE MobiCom (2004)
5. Tian, D., Georganas, N.: A node scheduling scheme for energy conservation in large wireless sensor networks. Wirel. Commun. Mob. Comput. J. **3**, 271–290 (2003)
6. Gui. C., Mohapatra, P.: Power conservation and quality of surveillance in target tracking sensor networks. In: Proceedings of ACM MOBICOM, Philadelphia, PA (2004)
7. Ren, S., Li, Q., Wang, H., Chen, X., Zhang, X.: Analyzing object tracking quality under probabilistic coverage in sensor networks. ACM Mob. Comput. Commun. Rev. **9**(1), 73–76 (2005)
8. Cao, Q., Abdelzaher, T., He, T., Stankovic, J.A.: Towards optimal sleep scheduling in sensor networks for rare event detection. In: Proceedings of the Fourth IEEE/ACM International Symposium on Information Processing in Sensor Networks, Los Angeles, CA (2005)
9. Kumar, S., Lai, T., Arora, A.: Barrier coverage with wireless sensors. In: ACM MOBICOM, Cologne, Germany (2005)
10. Rabbat, M.G., Nowak, R.D.: Decentralized source localization and tracking. In: Proceedings of the IEEE International Conference on Acoustics, Montreal, Canada, pp. 921–924 (2004)
11. Chen, W.P., Hou, J.C., Sha, L.: Dynamic clustering for acoutic target tracking in wireless sensor networks. In: Proceedings of the 11th IEEE International Conference on Network Protocols, pp. 284–294 (2003)
12. Yao, K., Hudson, R.E., Reed, C.W., Chen, D., Lorenzelli, F.: Blind beamforming on a randomly distributed sensor array system. IEEE J. Sel. Areas Commun. **16**, 1555–1567 (1998)
13. Gui, C., Mohapatra, P.: Virtual patrol: a new power conservative design for surveillance using sensor networks. In: Proceedings of the 4th International Symposium on Information Processing in Sensor Networks, pp. 246–253 (2005)
14. He, T., Krishnamurthy, S., Stankovic, J.A., Abdelzaher, T., Luo, L., Stoleru, R., Yan, T., Gu, L., Hui, J., Krogh, B.: Energy-efficient surveillance system using wireless sensor networks. In: Proceedings of the 2nd International Conference on Mobile Systems, Applications and Services, pp. 270–283 (2004)
15. Jeong, J., Gu, Y., He, T., Du, D.: VISA: virtual scanning algorithm for dynamic protection of road networks. In: IEEE INFOCOM, Rio de Janeiro, Brazil, pp. 927–935 (2009)
16. Jeong, J., Gu, Y., He, T., Du, D.H.C.: Virtual scanning algorithm for road network surveillance. IEEE Trans. Parallel Distrib. Syst. **21**(12), 1734–1749 (2010)

A Clique Based Asymmetric Rendezvous Scheme for Cognitive Radio Ad-Hoc Networks

Md Akbar Hossain[✉] and Nurul I. Sarkar

Department of IT and Software Engineering, Auckland University of Technology, Auckland, New Zealand
{akbar.hossain,nurul.sarkar}@aut.ac.nz

Abstract. Cognitive Radio (CR) is a promising technique to enhance the spectrum utilisation by enabling the CR users to opportunistic access the spectrum holes or channels. To exchange spectrum information most of the existing research have utilised a Common Control Channel (CCC). This results channel saturation, extreme transmission overhead of control information, and a point of vulnerability. To address this problem, Channel Hopping (CH) protocols have proposed for enabling Rendezvous (RDV). This paper presents a CH protocol based on clique system, called clique based channel hopping (CCH) for the purpose of RDV establishment. The proposed CCH is a role based blind RDV CH system where sender and receiver generates CH sequence based on h-clique and v-clique respectively. The CCH protocol satisfies the following requirements: (i) guaranteed RDV; (ii) no synchronisation; and (iii) symmetric and asymmetric channel model. Simulation results show that the proposed clique based channel hopping (CCH) scheme outperforms similar CH schemes in terms of average time-to-rendezvous (ATTR) and the degree of overlap in both symmetric and asymmetric channel scenario.

Keywords: Cognitive radio · Channel hopping · Rendezvous · Clique

1 Introduction

Cognitive radio ad-hoc networks (CRAHNs) is a dynamic multichannel environment whereby the channel status changes over time depending on surrounding radio users' activities. Hence, it is essential for a CR node to detect and identify its neighbours to initiate communication in CRAHNs, which corresponds to RDV. In a multi-channel wireless ad-hoc environment, RDV is the first key step for CR users to be able to communicate with each other. To achieve rendezvous, a CCC is often used as a common platform to exchange control information [1,2]. In CCC based protocol, two users tune the radio in the network wide or local common channel to exchange the control information. This common channel can be selected from unlicensed band or licensed band [3,4]. It is obvious that CCC

© ICST Institute for Computer Sciences, Social Informatics and Telecommunications Engineering 2018
P. H. J. Chong et al. (Eds.): SmartGIFT 2018, LNICST 245, pp. 268–277, 2018.
https://doi.org/10.1007/978-3-319-94965-9_27

based RDV schemes are very easy to deploy and efficient in centralized or coordinated network. Hence, in purely distributed ad-hoc networks (no coordination and no prior agreement), the performance of CCC based RDV protocols is limited to the factors; (i) single point failure, (ii) vulnerable to Primary User (PU) activity, (iii) channel saturation, and (iv) control channel jamming.

To overcome these performance instability, a channel hopping (CH) or parallel RDV is proposed in [5-7]. When a CR node wants to transmit a packet to its peer's, it switches from one channel to another by following a pre-defined [8] or random hopping sequence [9]. A desirable property of CH sequence scheme can guide any two users in the network to hop on the same channel in the same timeslot as soon as possible. Moreover, it should able to explore the temporal and spatial distribution of control channel in order to minimise the interference with license users. The authors in [10] proposed jump-stay based CH algorithm for both symmetric and asymmetric models without exploiting the time synchronization. However, ATTR in this case grows dramatically with the number of available channels as it does not have any preference to achieve rendezvous on a particular channel. Hence, the timeslots assignment on a given channel is not uniform and is a function of the LUs' activity [9]. The robustness against LUs' activity can be achieved by profiling (ranking) the available channels based on local channel sensing information. Only two CH sequences; AMRCC [9] and gQ-RDV [11] can be found in the existing literatures that consider channel quality to design CH sequence. Both of these protocols can guarantee rendezvous with symmetric channel information. However, there is no guarantee that they can achieve rendezvous on each available common channel with asymmetric channel information.

In this paper, we propose an asymmetric asynchronous channel hopping scheme by utilizing the clique system called clique based channel hopping (CCH). Our approach utilizes the mathematical concept of clique systems with rotation closure property to generate CH sequences that enable RDV on multiple channels. Moreover, the CCH scheme integrates the channel sensing information to assign the timeslots.

The rest of the paper is organized as follows: In Sect. 2, we presented the system model, channel activity model and followed by the RDV problem in CRAHNs. In Sect. 3, A detail description of channel hopping sequence generation is presented which includes grid formation, channel mapping and algorithmic analysis. In Sect. 4, simulation is performed to measure the performance of CCH protocol. Results show that, clique based CH sequence can guarantee to achieve RDV on number of available channels. We conclude our work in Sect. 5.

2 Preliminaries

2.1 System Model

We assume that there are N CR users in a single hoop CRAHNs that coexists with a number if PUs. Based on the time and space, CRs may observe same or different same of set of channels. It is considered that there are two types of channels available in the system; (i) licensed channel which is licensed to PUs and (ii) unlicensed channel which is open to all. A CR can operate in both linseed and unlicensed band. The available channels are indicated as $C = C_1, C_2, \ldots, C_M$. We also consider that each CR is capable to detect the spectrum hole and can operate on these channels. THe channel set of ith and jth CRs can be expressed as CR_i and CR_j respectively. The common channel set between user i and j can be represented as, $G_{i,j} = CR_i \bigcap CR_j$.

2.2 Channel Activity Model

A channel activity can be modeled as continuous time Markov chain (CTMC) where a channel can be any of these states (i) idle state (ii) PU state and (iii) CR state. A channel state can move from idle to PU or CR state if it is used by PU or CR respectively. The channel state can move form CR to PU if the PU reappear on the channel used by CR, however the channel state cannot move from PU to CR. The transition diagram is shown in Fig. 1.

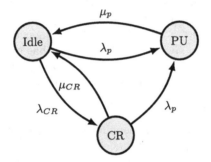

Fig. 1. Channel state transition diagram

Here the service request of both PUs and CR users modeled as a Poisson process with rate λ_P and λ_{CR} and is terminated with rate μ_p and μ_{CR}. Hence the transition matrix A can be written as:

$$\mathbf{A} = \begin{bmatrix} -(\lambda_p + \lambda_{CR}) & \lambda_p & \lambda_{CR} \\ \mu_p & -\mu_p & 0 \\ \mu_{CR} & \lambda_P & -(\lambda_P + \mu_{CR}) \end{bmatrix} \tag{1}$$

2.3 Problem Definition

In a multi-channel environment, the RDV problem can be described as a coordination problem where two suers have a set of channels to established communication with each other. However, RDV only achieved if both of the users are one the same channel at the same time. We assume that there are $P \geq 2$ CR users in the network, who share a set of available licensed channels, such as $X_i \subseteq X; (C \in X_0, X_1, X_2, \ldots, X_{N-1})$ (N is the number of available licensed channels and have labels that are the same for all CRs) that can be used for both control and data information exchange. Before initiating any data transmission, the intended nodes should first exchange control information between them to select common data channels. According to [10], nodes could have identical (i.e. $C_i = C_j$) or different available channel lists with at least one common channel, known as symmetric and asymmetric channel lists respectively. In CH-based solutions, the RDV process can be described as pairwise control channel establishment using sequences X and Y where X and Y are two CH sequences with period T, such as $(0, X[0]), (1, X[1]), \ldots, (T-1, X[T-1])$ and $(0, Y[0]), (1, Y[1]), \ldots, (T-1, Y[T-1])$. Hence, RDV sequences must have overlapped property in order to ensure any pair of nodes can establish communication, i.e. $\forall X, Y; |X \bigcap Y| \neq 0$.

3   Channel Hopping Sequence Design

The basic concept of this paper is inspired by the concept of clique. The clique is used in a social networks to identify the subset of people who all know each other. In this work, we use clique so that two users choose a set of channels from available channel list and jumps over the channel to achieve RDV. The size of the clique depends on the remaining energy of a CR node. In this section, we first provide a brief introduction of clique followed by CH sequence generation using clique.

Definition 1. Clique: Let us assume a finite universal set $U = \{0, 1, \ldots n-1\}$ of n elements, where n represents the cycle length. A clique C under universal set U is a collection of non-empty subsets of U, provided that it satisfies the intersection property: $\forall A, B \in Q : A \bigcap B \neq 0$.

For example $C = \{\{a, b\}, \{a, c\}, \{b, c\}\}$ is a clique under $U = \{a, b, c\}$. The basic idea of the CH scheme in CRAHNs is design hopping sequences so that two CRs can achieve rendezvous which is analogous to clique where two subsets have at least one intersection.

Definition 2. h-Clique: Given a finite universal set $U = \{0, 1, \ldots n-1\}$ of n elements, where n represents the cycle length. Let $l, 1 \leq l \leq \sqrt{(n)}$ and $m, 0 \leq m \leq n-1$ be two integers. An h-clique of l and m can be defined as follows:

$$h(m, l) = \left\{ \left(\left\lfloor \frac{\sqrt{n}}{l} i \right\rfloor \sqrt{n} + m + j \right) (mod\, n), \right.$$
$$\left. i = 0, \ldots, m-1; j = 0, \ldots, \sqrt{n} - 1 \right\} \tag{2}$$

For instance, when $n = 25$, for $m = 3$ and $l = 2$, the $h(3, 2) = \{3, 4, 5, 6, 7, 13, 14, 15, 16, 17\}$. Here m defines the starting position of the sequence such as $m = 3$ thus the sequence starts from 3. The purpose of l is to define the size of the sequence which is $l\sqrt{n}$. Figure 2(a) Shows an example of h-clique of h(3, 2).

Definition 3. v-Clique: Given a finite universal set $U = \{0, 1, \ldots n - 1\}$ of n elements, where n represents the cycle length. Let $l, 1 \leq l \leq \sqrt(n)$ and $p, 0 \leq p \leq n - 1$ be two integers. An v-clique of l and m can be defined as follows:

$$v(p, l) = \left\{ \left(\left\lfloor \frac{\sqrt{n}}{l} i \right\rfloor + p + j\sqrt{n} \right) (mod\, n), \right.$$

$$\left. i = 0, \ldots, m - 1; j = 0, \ldots, \sqrt{n} - 1 \right\} \tag{3}$$

For example, $n = 25$, for $p = 5$ and $l = 2$, the $v(5, 2) = \{0, 2, 5, 7, 10, 12, 15, 17, 20, 22\}$. Here p defines the starting portion of the element modulus n, such as $p = 5(mod\, n) = 0$ and 1 defines the number of columns as shown in Fig. 2(b).

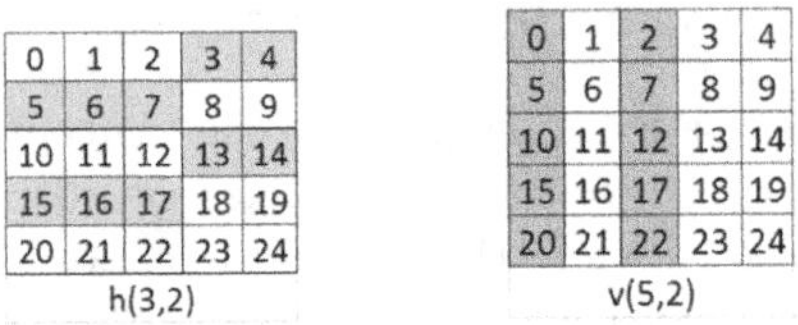

(a) h-clique of h(3,2)

(b) v-clique of v(5,2)

Fig. 2. CH sequence is generated by (a) h-clique and (b) v-clique for $n = 25$.

Upon joining in the network, a CR node will perform periodic sensing and ranked the available channel based on PU activity. Therefore, if a CR user has data to transmit it will construct a CH sequence based on h-clique and stay on each channel for T duration where T is the time required to exchange control information. The receiver is follow a CH sequence based on v-clique and visits the channel accordingly. The timeslot assignment for each channel is presented in Algorithm 1.

Theorem 1. Given two integers l_1 and l_2, $1 \leq (l_1, l_2) \leq \sqrt{n}$ and two random numbers m and p, $0 \leq (m, p) \leq n - 1$. The two users can achieve RDV within

$$\sqrt{n} \left(\left\lceil \frac{\sqrt{n}}{l_1} \right\rceil - 1 \right) + \left\lceil \frac{\sqrt{n}}{l_2} \right\rceil.$$

Proof: A sender generates a CH sequence based on h-clique which has $l_1 \times \sqrt{n}$ elements where the maximum distance between two successive elements can be written as $\sqrt{n} \left(\left\lceil \frac{\sqrt{n}}{l_1} \right\rceil - 1 \right) + 1$. Same as the receiver generates a CH sequence based on v-clique with $l_2 \times \sqrt{n}$ elements. The maximum distance between two CH sequences is the sum of the distance of sender and receiver CH sequence. We should subtract 1 from the summation to avoid a double count of the common element.

Algorithm 1. CCH Algorithm

Input: (i) Number of available channels, m;
      (ii) Transmission flag, $Flag_{Tx}$;
      (iii) Channel rank;
      (iv) Rescan period, T_{out};

Output: (i) Channel map CH_{map};
       (ii) Channel timeslots CH_{t_slots};

Begin

```
1:  while mod(t, T_out) = 0 do              15:      if T_new_slots < D then
2:      [Avail_CH]; {Available channel set}  16:          T_new_slots = [D, d] {d is dummy
3:      Rank([Avail_CH]) = [CH_List]                      variable to construct the grid}
4:      n = | Avail_CH |                     17:      else
5:      T_slots = n × n {Grid formation}     18:          T_new_slots = D
6:  end while                                19:      end if
7:  while packets arrive do                  20:      end for
8:      q ← q + 1                            21:  end while
9:      for i = 1 : n do                     22:  for j = 1 : n do
10:         m = randperm[0,n-1]              23:      p = randperm[0,n-1]
11:         construct h-clique h(m(i), l_1)  24:      construct  v(p(j), l_2)  {Receiver  CH
            {Sender CH sequence generation}           sequence generation}
12:         T_slots(h(m(i), l_1) = CH_list(i) 25:     T_slots(v(p(j), l_2) = CH_list(:, i)
13:         D = setdiff(T_slots, h(m, l_1))  26:  end for
14:         T_new_slots = (n - 1) × (n - 1)
```

End

A CH sequence in a time slotted architecture indicates the timeslots on which a node transmit or receive data to or from neighbors. Two CH sequence is called time synchronised if they start channel hopping at the same time as illustrated in Fig. 3, when $K = 0$. The value of K indicates the time lag between two CH sequence. K can be integer or fraction. If K is integer, we called it slot asynchronised otherwise it is fractional asynchronised. Both slot and fractional asynchronised scenarios are depict in Fig. 3. In fractional asynchronised case, rendezvous may not be achieved if the overlap duration is smaller than the time required to exchange control information. The performance of CH sequence depends on time synchronisation. Figure 3 shows that the degree of overlap between two CH sequences is significantly changes with time asynchronisation for different value of K. Consequently ATTR will also fluctuate due to the same. Hence, CH sequence that designed for time synchronize environment may not suit for time asynchronise environment.

Definition 4. Rotation: For a non-empty set X in a clique C under $U = \{0, \dots, n - 1\}$ and a non-negative integer $i \in \{1, 2, \dots, n - 1\}$, we define $rotate(x, i)$ or $R(X, i) = \{(j + i) \bmod n \mid j \in x\}$. A clique C is said to have the rotation closure property if $\forall X, X' \in C, i \in \{1, 2, \dots n - 1\} : X \bigcap X' \neq 0$.

For instance, the clique $C = \{\{a, b\}, \{a, c\}, \{b, c\}\}$ under $U = \{a, b, c\}$ satisfies rotation closure property. Thus, X and X' can still meet each other even though their clocks drifted. Thus, the rotation closure property holds for C. Two CR nodes adopting clique in C to denote their h-clique and v-clique can meet at the same time even if their clocks drifted.

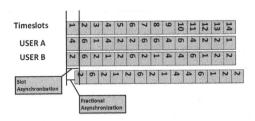

Fig. 3. Variation of degree of overlap in accordance with time synchronisation.

Theorem 2. Given two integers l_1 and l_2, $1 \le (l_1, l_2) \le \sqrt{n}$ and two random numbers m and p, $0 \le (m, p) \le n - 1$, then $R(h(m, l_1)) \bigcap R(v(p, l_2)) \ne 0$.

Proof: According to Definition 2 of h-clique, R(h) has at least $\sqrt{n}$ elements and any two successive elements must have distance either 1 of $n - \sqrt{n} + 1$. Same as Definition 3 of v-clique, R(v) has at least $\sqrt{n}$ elements and any two successive elements must have distance $\sqrt{n}$. Let y_i for $i = 0, \ldots, \sqrt{n} - 1$, be $\sqrt{n}$ elements of $R(v)$, then we can write $y_{i-1} \le y_i \le y_{i-1} + \sqrt{n} + 1$. Thus in order to proof the theorem we need to show that y_i is an element of $R(h)$. Let consider the smallest element of x in $R(h)$ which is larger than $y' \in R(v)$, then $y' + 1 \le x \le y' + n - \sqrt{n} + 2$ as any two elements in R(h) must have less than or equal to $n - \sqrt{n} + 1$. Thus $y' \in R(v)$, $x \le y' \le x + \sqrt{n} - 1$ which implies that y' is contained in R(h).

Theorem 3. Given two integers l_1 and l_2, $1 \le (l_1, l_2) \le \sqrt{n}$ and two random numbers m and p, $0 \le (m, p) \le n - 1$, then we have $| R(h(m, l_1)) \bigcap R(v(p, l_2)) | = l_1 \times l_2$.

Proof: From the Definition 2, $h(m, l_1)$ has a sequence of $\sqrt{n}$ contiguous and ascending elements and has at least one intersection with $v(p, 1)$ i.e. $|R(h(m, 1)) \bigcap R(v(p, 1))| = 1$. Now $h(m, l_1)$ has l_1 sequence which are disjoint and will have l_1 intersection with $v(p, 1)$. Same as $v(p, l_2)$ will have l_2 intersection with $h(m, 1)$. From this we can proved that $|R(h(m, l_1)) \bigcap R(v(p, l_2))| = l_1 \times l_2$. Figure 4 shows the graphical illustration of the Theorem 3. The first and second block of the Fig. 4, show the selection of timeslots according to $h(3, 2)$, and $v(5, 2)$, where $l_1 = 2$ and $l_2 = 2$. The third block of the Fig. 4 represents the timeslots where RDV is achieved. The no. of RDV is $l_1 \times l_2 = 2 \times 2 = 4$.

0	1	2	3	4
5	6	7	8	9
10	11	12	13	14
15	16	17	18	19
20	21	22	23	24

h(3,2)   v(5,2)   No. of RDV= 4

Fig. 4. Illustration of degree of overlap using clique channel hopping sequence.

4 Performance Evaluation

A MATLAB based simulation is used to evaluate the performance of the CCH protocol. A network with varying CR nodes is considered with the number of available channels ranging from 2 to 50 in a $800\,m \times 800\,m$ area, where each of the nodes had an equal transmission radius of $100\,m$. In this work both licensed and unlicensed channels were considered with equal priority. CR nodes were considered asynchronous, which was implemented by imposing a random delay at network initialization. During the simulation, the CR nodes may synchronise themselves after achieving RDV. Each CR node starts with spectrum sensing. The sensing duration was set at $25\,ms$/channel and $\leq 1\,ms$/channel for fine and fast sensing respectively [12]. Fast sensing is performed by selecting samples of the PU Poisson traffic within its sensing period and later on performing fine sensing before jumping into a channel. The ranking table of CR nodes is based on channel availability and channel activity observed locally by a CR user. It is assumed that, if a packet arrives during the spectrum sensing or handshaking process, it is enqueued and remains in the queue till RDV is achieved. In this chapter, collision among control or handshake packets is not considered; however, in case of collision between a CR user packet and a PU packet, the CR user packet is dropped instead of being retransmitted after a backoff.

All the results presented in this chapter are averaged over 10000 iterations [13]. Each PU is randomly assigned a channel when a new packet needs to be transmitted and packet arrivals follow the Poisson distribution with exponentially distributed inter-arrival times.

4.1 Impact of Number of Channels

Figure 5 shows the impact of number of channels on the network performance which includes ATTR and RDV success rate. We have considered maximum 50 channels that are available for CR users to achieve RDV. Figure 5(a) shows that the ATTR is increasing with number of channels for all protocols including our proposed CCH as the CH sequence length is proportional to number of channels in the available channel set. However, the prosed CCH protocol exhibits superior performance compare to other protocols as it facilitates higher number of RDV in each CH sequence cycle. For gQ-RDV, the number of RDV on each cycle is fixed i.e. 2. In JS, the ATTR is at least three (3) times of the number of available channels. The ATTR of CCH is 20 for 30 channels, while it is 38, 70, and 117 for JS, gQ-RDV and basic AMRCC respectively. The higher the ATTR, the lower the RDV success rate which is shown in Fig. 5(b). RDV success rate is a design issue of the CH sequence design. In CCH, we have a very clear control by setting the two parameters l_1 and l_2. In JS, CR users have a guaranteed RDV on each channel however there is no control how many RDV can be achieved in each CH cycle. For instance, when there are 25 channels in the available channel set, the RDV success rate is 87% for CCH and 73%, 28%, and 3.5% for JS, gQ-RDV, and basic AMRCC.

4.2   Impact of Asymmetry

Due to temporal and spatial changes in radio environment, CR users may observe different set of channels and have different cardinality of the available channel set. The degree os asymmetry can be defined as $\alpha = \frac{|C_x|}{|C_y|}$, where $|C_x|$ and $|C_y|$ are the cardinality of the available channel set of user x and y. As basic AMRCC doesn't support the asymmetric channel, we have excluded it from this analysis. We have considered two values for α and Fig. 6(a) illustrate the ATTR of CCH, JS, and gQ-RDV for $\alpha = 0.4$ and $\alpha = 0.8$. For simplicity we considered a fixed number of common channels between users. At each run, the common channels are selected randomly from the available channel set. A similar trend of ATTR is observed for all the protocols. The CCH outperforms than other two protocols. For instance, when the number of channel is 35 with degree of asymmetry is 0.8, the ATTR for JS and gQ-RDV is 14.44% and 131% higher than that of CCH. Interestingly, the performance gap between different protocols is increased with the degree of asymmetry. The same behaviour can also be observed for RDV success which is shown in Fig. 6(b).

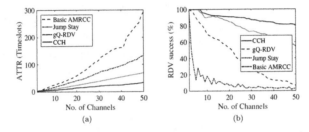

Fig. 5. Comparison of ATTR and degree of overlap with increasing number of available channels: (a) ATTR (b) Rendezvous success.

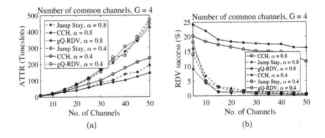

Fig. 6. Comparison of ATTR and degree of overlap when two users experience asymmetric set of channels: (a) ATTR (b) Rendezvous success

5   Conclusion

In this paper, a clique based asymmetric CH sequence is proposed to solve the RDV problem. Using a clique based system grantees that the nodes would meet within a bounded time. This method satisfies the both asynchornous and asymmetric blind RDV. We have performed algorithmic analysis in terms of ATTR and RDV success rate. Simulation results have shown that our CH scheme is highly resilient to dynamic spectrum allocation in CRAHNs. The extension of CCH for multiuser multihop communication would consider as our future works.

References

1. Yoo, S., Nan, H., Hyon, T.: DCR-MAC: distributed cognitive radio MAC protocol for wireless ad hoc networks. Wirel. Commun. Mob. Comput. **9**(5), 631–653 (2009)
2. Jia, J., Zhang, Q., Shen, X.: HC-MAC: a hardware-constrained cognitive MAC for efficient spectrum management. IEEE J. Sel. Areas Commun. **26**(1), 106–117 (2008)
3. Yin, S., Chen, D., Zhang, Q., Li, S.: Prediction-based throughput optimization for dynamic spectrum access. IEEE Trans. Veh. Technol. **60**(3), 1284–1289 (2011)
4. Ma, L., Han, X., Shen, C.-C.: Dynamic open spectrum sharing MAC protocol for wireless ad hoc networks. In: IEEE International Symposium on New Frontiers in Dynamic Spectrum Access Networks, pp. 203–213, November 2005
5. Bahl, P., Chandra, R., Dunagan, J.: SSCH: slotted seeded channel hopping for capacity improvement in IEEE 802.11 ad-hoc wireless networks. In: Proceedings of the 10th Annual International Conference on Mobile Computing and Networking, pp. 216–230. ACM (2004)
6. So, W., Walrand, J., Mo, J., et al.: McMAC: a parallel rendezvous multi-channel MAC protocol. In: IEEE Wireless Communications and Networking Conference, pp. 334–339. IEEE (2007)
7. Liu, H., Lin, Z., Chu, X., Leung, Y.-W.: Ring-walk based channel-hopping algorithms with guaranteed rendezvous for cognitive radio networks. In: IEEE/ACM International Conference on Cyber, Physical and Social Computing, pp. 755–760, December 2010
8. Shin, J., Yang, D., Kim, C.: A channel rendezvous scheme for cognitive radio networks. IEEE Commun. Lett. **14**(10), 954–956 (2010)
9. Cormio, C., Chowdhury, K.R.: Common control channel design for cognitive radio wireless ad hoc networks using adaptive frequency hopping. Ad Hoc Netw. **8**(4), 430–438 (2010). https://doi.org/10.1016/j.adhoc.2009.10.004
10. Liu, H., Lin, Z., Chu, X., Leung, Y.-W.: Jump-stay rendezvous algorithm for cognitive radio networks. IEEE Trans. Parallel Distrib. Syst. **23**(10), 1867–1881 (2012)
11. Romaszko, S., Mähönen, P.: Quorum-based channel allocation with asymmetric channel view in cognitive radio networks. In: Proceedings of the 6th ACM Workshop on Performance Monitoring and Measurement of Heterogeneous Wireless and Wired Networks, ser. PM2HW2N 2011, pp. 67–74. ACM, New York (2011). https://doi.org/10.1145/2069087.2069097
12. Cordeiro, C., Challapali, K., Birru, D., Sai Shankar, N.: IEEE 802.22: the first worldwide wireless standard based on cognitive radios. In: First IEEE International Symposium on New Frontiers in Dynamic Spectrum Access Networks, pp. 328–337, November 2005
13. Chuang, I., Wu, H.-Y., Lee, K.-R., Kuo, Y.-H.: Alternate hop-and-wait channel rendezvous method for cognitive radio networks. In: IEEE INFOCOM proceedings, pp. 746–754, April 2013

The Blockchain Marketplace as the Fifth Type of Electricity Market

Yueqiang Xu[1(✉)], Petri Ahokangas[1], Seppo Yrjölä[2],
and Timo Koivumäki[1]

[1] Oulu Business School, University of Oulu,
Finland Pentti Kaiteran katu 1, P.O. Box 4600, 90014 Oulu, Finland
{yueqiang.xu,petri.ahokangas,timo.koivumaki}@oulu.fi
[2] Nokia, Kaapelitie 4, 90620 Oulu, Finland
seppo.yrjola@nokia.com

Abstract. This paper tackles today's unprecedented challenges of enabling and stimulating multiple energy stakeholders to have a more active participation in the smart grid electricity market. The research extends the existing four archetypes of orchestrator-driven business models for the electricity market and proposes a fifth type of electricity market, the Blockchain Marketplace. The key novelty of the paper is to expand the electricity market architecture and design from centralization and pseudo-decentralization to full decentralization, enabled by the blockchain. The study not only broadens the smart grid and electricity market literature but also contributes to the theoretical development of the business model and organization study domains with a systemic approach.

Keywords: Electricity market · Blockchain · Business model
Smart grid

1 Introduction

The energy industry used to have a simple business model. Fully integrated electric companies used to be the center of the industry, building transmission and distribution networks for the constantly growing demand. Energy utilities decided when and where to invest and build generation capacities; they decided how to maintain the system in balance, acting as a centralized controlling entity.

However, the world is changing and electric systems have been undergoing significant changes [1]. The current electrification state of the world is at 85.3% [2]. The growing energy demand and dependence on fossil fuel has become a global issue [3]. Simultaneously, the integration of large volumes of distributed energy resources (DERs) has posed unprecedented challenges to maintaining the balance of generation and demand as well as planning and operations of concurrent electric infrastructure [4].

It is urged by [5] that a global technological revolution is changing the power balance between consumers and centralized utilities. The increasing growth in DER is moving the power balance from integrated utilities to the demand side, where consumers have control over a more sustainable, more local, and more resilient energy system. Numerous studies address the need to move this revolution into the mainstream and create a new model or design for the energy market that puts consumers in charge of a co-created energy future [6, 7].

The European Union (EU) just announced the ambitious "Clean Energy for All Europeans" package [8], demanding "an increase in the energy efficiency target from 27% to 30%, a cut in emissions by 40%, and a goal of 27% renewables in final consumption, all by 2030". Thus, a key question is how to enable energy stakeholders (e.g., consumers, prosumers, DERs, utilities) to have more active participation in the energy market, facilitating the evolution of energy system to save the planet while creating more value for the market participants.

The blockchain, developed first for the Bitcoin cryptocurrency [9], is a decentralized transaction and data management technology and a distributed database solution maintaining growing data records that are confirmed by the nodes participating in it [10]. Industry-specific blockchain use cases are being identified in different fields, such as finance [5], telecommunication and spectrum sharing [11], and the Internet of Things (IoT) [12]. The research of [13] focuses on the blockchain's application in energy, identifying use cases like solar trading in the US, energy exchange in Austria, and the billing process for autonomous electric vehicle charging stations in Germany.

This paper aims at contributing to the research on new electricity market design and business models as follows: (1) the paper studies the most recent theoretical development of the business model, resource configuration, and organization design literature for the systemic design of the electricity market in the digital age; (2) the research elevates the extant four archetypes of resource configurations (orchestrator-driven, from centralization to pseudo-decentralization) to a systemic logic (system-driven, full decentralization) that was not there in the previous literature; (3) from the business model perspective, the study proposes a fully decentralized electricity market design enabled by the blockchain or the so-called "Blockchain Marketplace" as opposed to the centralized and pseudo-decentralized electricity market design in the existing literature.

The paper is composed as follows: In Sect. 2, the paper provides a characterization of the four established resource configuration business model archetypes. Section 3 provides a concise description of the blockchain technology. Section 4 describes the research methodology and data collection. Section 5 constructs a blockchain-based business model archetype for the future energy market by first discussing the identifiable applications of the four archetypes in the energy market as the context and then proposing the concept of a fifth electricity marketplace. Section 6 provides a concluding discussion on implications and limitations and a recommendation for future studies.

2 The Four Archetypes of the Business Model

Business model research has expanded during the last decades [14]. The business model has been studied as a system/collection of interdependent components such as resources and competence, internal and external organizational structures [15], customer value proposition [16], and cost and revenue structure [17]. Only recently, a converging conceptualization, incorporating three key processes for business models that connect them to the context, opportunity processes (exploration, exploitation) [18], value processes (creation, capture) [19], and advantage processes [7], has emerged in the scientific community.

Other business model discussion streams have focused on the discovery or design of successful business models. The literature shows that viability (robust performance [20]), sustainability (technical-, social- and environmental-oriented [21]), and scalability (scalable deployment capability [22] and profitable growth [23]) are essential to business model success as "a better way than the existing alternatives" [24].

In the pervasive digital age, the scope of resources that a company can utilize and access has expanded. A holistic approach is required to enhance value creation and capture from [25]'s "added-value" strategy to the latest business model discussions on value centricity [26] and systemic value [7] for digital business models. Building on these latest approaches and resource orchestration [27], this paper first opens up the four archetypes of orchestrator-driven business models before diving into the fifth archetype in Sect. 4:

- **The first archetype—"company as an integrator":** The focal orchestrating firm (O) transforms resources to create value for customers. This has been the predominant type of resource configuration for traditional companies like manufacturers, studied in the light of established theories [27].
- **The second archetype—"company as a collaborator":** The orchestrating firm (O) collaborates with partners who have complementary resources as a value-creating resource configuration. This archetype is recognized in strategic alliance [28] and ecosystem studies [29].
- **The third archetype—"company as a transaction enabler":** This is associated with the platform business model, meaning that broader and easier access to resources allows the orchestrating firm (O) to build two or multi-sided markets to match resources and needs.
- **The last archetype—"company as a bridge":** This shows that the proliferation of virtual resources (such as data) creates the opportunity for an orchestrating firm (O) to bridge certain groups of market participants that have not been previously connected, based on the data and benefiting from bridging unconnected needs, such as Google's advertising model [26].

3 Blockchain and Smart Contract

According to [30], the idea to collaboratively consume, share, and decentralize resources or assets among different peers can be seen in various concepts such as the Sharing Economy [31], Collaborative Consumption [32], and the Peer-to-Peer (P2P) Economy [33]. Blockchain technology is identified as the enabler for such a fully decentralized system [30].

Blockchain is a general-purpose decentralized transaction and data management technology, developed first for the Bitcoin cryptocurrency [9], with the ability to track transactions, settle trade deals, and enforce contracts across a wide range of digital assets which in turn can represent currency, IP, data, contracts or physical assets [34].

In practice, a blockchain is a distributed database solution maintaining a continuously growing list of data records that are confirmed by the nodes participating in it. The data is recorded in a public ledger, including information about every transaction completed. A blockchain network is a peer-to-peer (P2P) network that does not require a third-party organization in the middle. As no central server or intermediary is in place, a consensus mechanism is needed for ensuring the coherency of data between the nodes. There are several consensus mechanisms under discussion, e.g., [35, 36]. Furthermore, in the blockchain, the utilization of cryptography enables authoritativeness behind all interactions [37]. Information about every completed transaction is shared and available to all nodes, which makes the system more transparent than centralized solutions [38]. The extant literature claims that the blockchain is embodied in a combination of existing technologies, including peer-to-peer networks, cryptographic algorithms, distributed data storage, and decentralized consensus mechanisms [30].

Smart contracts (SC) operate as autonomous actors with self-executing scripts that reside on the blockchain, enabling general-purpose computations occurring on the chain to be fully predictable [37]. The SC concept was introduced in 1994 [39], defined as a computerized transaction protocol that executes the terms of a contract. The SC code and the cryptographically verifiable trace of the SC's operations can be inspected by all the network participants. SC enables the automation of complex multi-step processes and proper, distributed, heavily automated workflows [37]. They have many applications in different domains, enabling, e.g., decentralized applications like voting, auctions, lottery, escrow systems, crowdfunding, and micropayments [40].

It is suggested by [41] that the blockchain technology is poised to improve the smart grids that incorporate communication technology and sensors. This can range from super grids that connect large-scale energy systems (e.g., storage) to microgrids that are designed for connecting DERs. In fact, a number of blockchain energy initiatives are emerging globally, such as the Brooklyn Microgrid [42].

The antecedents of blockchain-enabled electricity trading and marketplace can be found in both the conceptual and empirical realms. For instance, peer energy trading is one of the highly promising areas for the Blockchain Marketplace [42]. Conceptually, a case of decentralized sharing in photovoltaic (PV) generation is proposed by [41]. The conceptual use case investigates the autonomous optimization and energy trading among different systems (including heating, cooling, hot water storage, and energy storage), which resembles a localized machine-to-machine electricity market.

The Brooklyn Microgrid is an empirical example where household residents trade energy among themselves. This blockchain platform provides the technical infrastructure for the local electricity market. Prosumers and consumers can submit, buy, or sell electricity orders to the market through the pre-defined market mechanism [42].

GrünStromJeton is another case studied in the European Commission's report [43]. GrünStromJeton provides an index that indicates the relative production of energy from alternative renewables during the next 36 h. The system monitors and records the energy consumption of the customers and rewards consumers when they use renewable energy sources. This is a trading mechanism between GrünStromJeton's digital system and the actual consumers.

4   Research Methodology and Data Collection

This study adopts the action research methodology within techno-social innovation research as part of a major EU energy innovation project on a P2P trading platform that enables and supports a decentralized energy market design and the P2P energy exchange of smart grids. According to [44], the action research methodology leads to producing scientific knowledge that can serve the actions, which enables the formalization and contextualization of models and tools, facilitating the production of new knowledge and enabling organization change.

The research was conducted in two steps: First, the study embarks on a systematic analysis of 50 innovative business cases with various types of business models in the energy and smart grid industry. The data was collected from the European Commission's BRIDGE initiative, uniting 31 major European smart grid and energy storage projects. The collected data is in the form of business model examples, which are contributed by international energy experts, energy companies, regulators, and research organizations. The comprehensive collection of business model data enables a thorough analysis of business model archetypes in the energy industry, avoiding common selection bias [45].

The second step utilized the business model design framework used by [26] on resource configurations for digital business. This study adopted a systemic and value co-creation centric perspective that considered the needs (N), the resources (R), and the created value (V-C) of all value co-creators in the energy industry. Such an approach is grounded in the literature of resource orchestration [27] and business model design [20].

5   The Fifth Archetype of the Business Model for the Electricity Market

In this section, the **four archetypes of orchestrator-driven business models in the electricity market** are first presented as four prototypes (Fig. 1).

Prototype A: The Centralized Utility Model. This is the traditional utility business model which assembles a "company as an integrator". It has the simplest resource configuration, where a traditional integrated utility (orchestrator) converts the generation fuel (e.g., coal, natural gas) as resources (R_{CU}) to address the consumer's electricity consumption as demand (N_1). Consumers contribute to the business need utilities (N_{CU}) as revenue with financial resources (payment for an electricity bill) (R_1). It is argued by [6] that this prototype makes very little space for the growth of DERs and local demand services and gives poor support for energy efficiency.

Prototype B: The Disintegrated Retailer Model. A disintegrated retailer model is defined by [6] as an organization that does not own generation assets (such as power plants), instead partnering with one or more generators and using its own brand. It is a common business model for electricity retailers in liberalized markets. As a Prototype B utility, the orchestrating retailer company collaborates with a partner with generation assets ($V\text{-}C_2$) to supply and service the needs of consumers (N_1). The

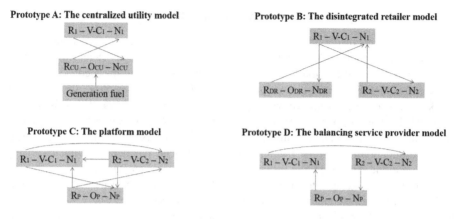

Fig. 1. The four archetypes/prototypes of business models in the energy market.

resources to meet the consumption (R_{DR} and R_2) are not solely from the disintegrated retailer but are contributed by its partners. Thus, the role of a Prototype B company is not a resource transformer as in Prototype A, but a collaborator who engages another "complementor" (V-C_2) [29] to create value for energy consumers (V-C_1).

Prototype C: The Platform Model. The platform research was pioneered by [46]. Cambridge researchers [47] initiated the discussion of the platform business model in the energy market. A platform operator brings together groups of users and providers of products and services, mediating their interaction and matching needs. A key feature of the platform market is the existence of the network effect: the value of the platform changes with respect to the participation rates on the same side and the cross side [47]. In the energy market, the platform operator (orchestrator) contributes resources (R_P) to enable interactions and matchmaking between two groups of value co-creators (such as consumer and prosumer) whose needs (N_1 and N_2) can be matched by each other's resources (R_2 and R_1). Thus, the platform operator facilitates energy trading between consumers and prosumers as well as among several prosumer groups.

Prototype D: The Balancing Service Provider Model. Balancing market is defined by [48] as the institutional arrangement that establishes market-based balance management in an unbundled electricity market. The business model for the balancing service operator can be a virtual power plant (VPP) [44] or a local balancing unit [6]. A balancing service provider adopts a Prototype D business model, using its resources (RBSP) such as a digital solution to provide energy efficiency services for the needs of consumers (N_1). As an orchestrating entity, the balancing service provider utilizes consumption data and patterns collected from the consumer (R_1) to address the needs of another group (N_2), such as the distribution network operators (DSOs). Thus, the balancing service provider enables DSOs to leverage the consumption data and behaviors controlled by the consumers to balance the electricity network.

5.1   The Fifth Type of Electricity Market, the Blockchain Marketplace

In a blockchain-enabled market, the blockchain operates as a chronological, immutable, and trusted data storage. The smart contracts generated by the blockchain can automate offer testing and modification based on the parameters tuned in the feedback loop.

Based on the above discussion, Fig. 2 shows the simplest form of the blockchain-enabled marketplace without the need of an orchestrator. The existence of the blockchain and smart contract allows any participant in the energy market, such as participant 1, to match its N_1 with the R_2 of participant 2 (such as the case of energy and flexibility trading), while the R_1 (e.g., the financial payment) is directed to N_2 without an orchestrator standing between the direct value co-creation and co-capture. In contrast, with the orchestrator-driven models (Fig. 3), a portion of the value flows out of the direct value co-creation and is captured by the orchestrator. Theoretically, in the Blockchain Marketplace, there is no value flowing out of the direct value co-creation and market participants are better off with more value accrued and shared.

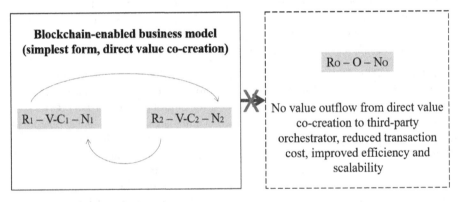

Fig. 2. The simplest form of a blockchain-enabled marketplace.

According to [49], transaction costs are fundamentally interrelated with distrust. As a micro process of the business model [26], trust as one of the key characteristics of the blockchain facilitates the resource crowdsourcing process by providing lower transaction costs. The costs of verification and networking can be radically reduced through the blockchain [34], while the parts of the transaction that concern negotiating, establishing, and enforcing the transaction may be automated as smart contracts operating on behalf of value co-creators, making and accepting tenders, matching needs and resources. Instead of a centralized orchestrator, the Blockchain Marketplace can automate large amounts of decentralized transactions, reducing transaction costs, increasing direct value co-creation flows, and improving efficiency and scalability with no need for a third-party intermediary, as shown in Fig. 3.

Based on a trusted data set provided by the blockchain, prospecting and sorting algorithms can be used for further tuning the business processes in a Blockchain Marketplace for the energy industry. To mitigate general privacy risk of the blockchain

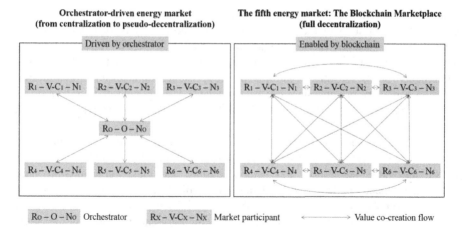

Fig. 3. Comparing the orchestrator-driven energy market and the blockchain marketplace.

technology related to gathering extensive data sets, new secure privacy-preserving encryption methods are being developed, such as [50, 51].

As part of the process, the blockchain can also facilitate the novel grafting of resource combinations and configurations, streamlining and fine-tuning the smart contract parameters controlling the relationship between energy market participants, and combining data analysis tools with the digital blockchain platform to enrich novel complementarity created in the grafting process.

6   Concluding Remarks

This paper uses the resource configuration approach to propose the fully decentralized business model archetype for a blockchain-enabled electricity market, pointing out that despite the continuous liberalization effort of the regulation, the electricity market design and business model remains a centralized scheme or a pseudo-decentralization at its best. It identifies how the blockchain as a new technology or technical development can affect the market design of electricity, contributing the business model perspective to the extant energy and electricity market literature.

The **theoretical contributions** of the study are as follow: The resource configuration approach explains how a company can create and capture value through strategically configuring the resource [27, 52], which is further embodied in business model studies with the latest classification of four business model archetypes [26]. This study identifies that the four archetypes are only sufficient to depict orchestrator-driven business models, since all the archetypes require an orchestrating entity to enable or facilitate the value co-creation while extracting part of the value to meet its own needs (such as revenue and profit).

To tackle the aforementioned (research) gap, this study looks into the fully decentralized business model concepts and proposes the Blockchain Marketplace as a fifth business model archetype for the energy market, and the only archetype that is not orchestrator-driven and promising full autonomy for the market participants. The theoretical contribution of the paper is not limited to energy market design; it also contributes to management and organization studies in general through systemic thinking about how resources are configured, how value is created, and how market participants' needs are met, regardless of the existence of a resource-orchestrating entity. An industry or ecosystem can be formulated from pseudo-decentralization (as managed by an orchestrating entity) to full decentralization (as enabled by the blockchain).

Regarding the **empirical contributions**, the blockchain is expected to change and affect the centralized legacy systems with full decentralization, enabling microgrids, DERs, renewable integration, P2P energy trading and higher consumer/prosumer engagement, managing less predictable and more volatile renewable power sources in the future. The empirical goal is to help the energy industry to re-think value creation, breaking the boundaries and constraints of contemporary energy market design logic, enabling the discovery of new design patterns for innovative resource configuration for a future-oriented electricity market.

This study is limited and focused on the conceptualization of the Blockchain Marketplace. The micro processes of the business model development are only briefly touched upon in the paper. In the future, the micro processes for the development of this new business model archetype need to be further studied and discussed to shed light on how the fifth type of electricity market, the Blockchain Marketplace, functions. The further development of this new archetype through simulation and quantitative methods is recommended for future research.

It is noteworthy that this study does not suggest that the blockchain technology can tackle all the issues of today's energy system, but rather it facilitates and enhances a more meaningful archetypical design for the energy industry in the digital era.

References

1. Wang, J., Conejo, A.J., Wang, C., Yan, J.: Smart grids, renewable energy integration, and climate change mitigation - future electric energy systems. Appl. Energy **96**, 1–3 (2012)
2. World Bank: Access to Electricity (Percentage of Population). http://data.worldbank.org/indicator/EG.ELC.ACCS.ZS. Accessed 22 Nov 2017
3. Mohamed, M.A., Eltamaly, A.M., Farh, H.M., Alolah, A.I.: Energy management and renewable energy integration in smart grid system. In: 2015 IEEE International Conference on Smart Energy Grid Engineering, pp. 1–6 (2015)
4. Nistor, S., Wu, J., Sooriyabandara, M., Ekanayake, J.: Capability of smart appliances to provide reserve services. Appl. Energy **138**, 590–597 (2015)
5. Hasse, F., von Perfall, A., Hillebrand, T., Smole, E., Lay, L., Charlet, M.: Blockchain - an opportunity for energy producers and consumers? (2016)
6. Hall, S., Roelich, K.: Business model innovation in electricity supply markets: the role of complex value in the United Kingdom. Energy Policy **92**, 286–298 (2016)
7. Xu, Y., Ahokangas, P., Reuter, E.: EaaS: electricity as a service? In: 24th Nordic Academy of Management Conference, pp. 1–22. NFF, Bodø (2017)
8. European Commission: Clean Energy for All Europeans. Brussels (2016)

9. Nakamoto, S.: Bitcoin: A Peer-to-Peer Electronic Cash System (2008). https://bitcoin.org/bitcoin.pdf. Accessed 27 Nov 2017

10. Antonopoulos, A.M.: Mastering Bitcoin: Unlocking Digital Cryptocurrencies. O'Reilly Media Inc., Sebastopol (2014)

11. Yrjölä, S.: Analysis of blockchain use cases in the citizens broadband radio service spectrum sharing concept. In: Marques, P., et al. (eds.) CrownCom 2017. LNICST, vol. 228, pp. 128–139. Springer, Cham (2018). https://doi.org/10.1007/978-3-319-76207-4_11

12. Zhang, Y., Wen, J.: The IoT electric business model: using blockchain technology for the internet of things. Peer-to-Peer Netw. Appl. **10**, 983–994 (2017)

13. Basden, J., Cottrell, M.: How utilities are using blockchain to modernize the grid. Harv. Bus. Rev. 2–5 (2017)

14. Mazhelis, O., Warma, H., Leminen, S., Ahokangas, P., Pussinen, P., Rajahonka, M., Siuruainen, R., Okkonen, H., Shveykovskiy, A., Myllykoski, J.: Internet-of-Things Market, Value Networks, and Business Models: State of the Art Report, Jyväskylä (2013)

15. Demil, B., Lecocq, X.: Business model evolution: in search of dynamic consistency. Long Range Plann. **43**, 227–246 (2010)

16. Johnson, M., Christensen, C.M., Kagermann, H.: Reinventing Your Business Model. Harv. Bus. Rev. 50–60 (2008)

17. Osterwalder, A., Pigneur, Y.: Business Model Generation: A Handbook for Visionaries, Game Changers, and Challengers. Wiley, Hoboken (2010)

18. Chesbrough, H.: Business model innovation: opportunities and barriers. Long Range Plann. **43**, 354–363 (2010)

19. Zott, C., Amit, R.: The business model: a theoretically anchored robust construct for strategic analysis. Strateg. Organ. 1–20 (2013)

20. Amit, R., Zott, C.: Crafting business architecture: the antecedents of business model design. Strateg. Entrep. J. **9**, 331–350 (2015)

21. Biloslavo, R., Bagnoli, C., Edgar, D.: An eco-critical perspective on business models: the value triangle as an approach to closing the sustainability gap. J. Clean. Prod. **174**, 746–762 (2017)

22. Moqaddamerad, S., Xu, Y., Iivari, M., Ahokangas, P.: Business models based on co-opetition in a hyper-connected era: the case of 5G-enabled smart grids. In: Afsarmanesh, H., Camarinha-Matos, L.M., Lucas Soares, A. (eds.) PRO-VE 2016. IAICT, vol. 480, pp. 559–568. Springer, Cham (2016). https://doi.org/10.1007/978-3-319-45390-3_47

23. Holzner, L., Bohnsack, R.: Business model adaptation mechanisms in the internationalization process: the case of energy firms. In: 43rd European International Business Academy Conference, Milan, pp. 1–39 (2017)

24. Magretta, J.: Why business models matter. Harv. Bus. Rev. **80**, 86–87 (2002)

25. Brandenburger, A.M., Stuart, H.W.J.: Value based business strategy. J. Econ. Manag. Strateg. **5**, 5–24 (1996)

26. Amit, R., Han, X.: Value creation through novel resource configurations in a digitally enabled world. Strateg. Entrep. J. **11**, 228–242 (2017)

27. Sirmon, D.G., Hitt, M.A., Ireland, R.D., Gilbert, B.A.: Resource orchestration to create competitive advantage. J. Manag. **37**, 1390–1412 (2011)

28. Wassmer, U., Dussauge, P.: Network resource stocks and flows: how do alliance portfolios affect the value of new alliance formations? Strateg. Manag. J. **33**, 871–883 (2012)

29. Adner, R., Kapoor, R.: Value creation in innovation ecosystems: how the structure of technological interdependence affects firm performance in new technology generations. Strateg. Manag. J. **31**, 306–333 (2010)

30. Löbbers, J., von Hoffen, M., Becker, J.: Business development in the sharing economy: a business model generation framework. In: 2017 IEEE 19th Conference on Business Informatics (CBI), pp. 237–246. IEEE, Thessaloniki (2017)
31. Malhotra, A., Van Alstyne, M.: The dark side of the sharing economy … and how to lighten it. Commun. ACM **57**, 24–27 (2014)
32. Botsman, R., Rogers, R.: What's Mine Is Yours - How Collaborative Consumption Is Changing the Way We Live. Harper Collins Inc., New York (2010)
33. Botsman, R.: The sharing economy lacks a shared definition. http://www.fastcoexist.com/3022028/the-sharing-economy-lacks-a-shared-definition#1. Accessed 22 Nov 2017
34. Catalini, C., Gans, J.S.: Some simple economics of the blockchain. SSRN Electron. J. (2017)
35. Vukolić, M.: The quest for scalable blockchain fabric: proof-of-work vs. BFT replication. In: Camenisch, J., Kesdoğan, D. (eds.) iNetSec 2015. LNCS, vol. 9591, pp. 112–125. Springer, Cham (2016). https://doi.org/10.1007/978-3-319-39028-4_9
36. Kiayias, A., Russell, A., David, B., Oliynykov, R.: A Provably Secure Proof-of-Stake Blockchain Protocol (2016)
37. Christidis, K., Devetsikiotis, M.: Blockchains and smart contracts for the internet of things. IEEE Access **4**, 2292–2303 (2016)
38. Yli-Huumo, J., Ko, D., Choi, S., Park, S., Smolander, K.: Where is current research on blockchain technology? A systematic review. PLoS ONE **11**, 1–27 (2016)
39. Szabo, N.: Smart Contracts. https://archive.is/X3lR2. Accessed 22 Mar 2018
40. Baliga, A.: The Blockchain Landscape Office of the CTO (2016)
41. Meisel, M., Fotiadis, L., Wilker, S., Treytl, A., Sauter, T.: Blockchain applications in microgrids: an overview of current projects and concepts Andrija. In: IECON 2017 - 43rd Annual Conference IEEE Industrial Electronics Society, pp. 6153–6158 (2017)
42. Mengelkamp, E., Gärttner, J., Rock, K., Kessler, S., Orsini, L., Weinhardt, C.: Designing microgrid energy markets: a case study: the Brooklyn Microgrid. Appl. Energy **210**, 870–880 (2018)
43. Ioannis, K., Raimondo, G., Dimitrios, G., Gioia Rosanna, D., Georgios, K., Gary, S., Ricardo, N., Igor, N.-F.: Blockchain in Energy Communities. Brussels (2017)
44. Bahari, N., Maniak, R., Fernandez, V.: Ecosystem business model design. In: XXIVe Conférence Internationale de Management Stratégique, pp. 1–18. AIMS, Paris (2015)
45. Collier, D., Mahoney, J.: Insights and pitfalls: selection bias in qualitative research. World Polit. **49**, 56–91 (1996)
46. Bakos, Y.: The emerging role of electronic marketplaces on the Internet. Commun. ACM **41**, 35–42 (1998)
47. Weiller, C.M., Pollitt, M.G.: Cambridge Working Platform Markets and Energy Services, Cambridge (2013)
48. van der Veen, R.A.C., Hakvoort, R.A.: The electricity balancing market: exploring the design challenge. Util. Policy **43**, 186–194 (2016)
49. Cai, R.: Trust and Transaction Costs in Industrial Districts (2004). http://scholar.lib.vt.edu/theses/available/etd-05222004-232528/. Accessed 27 Nov 2017
50. Zyskind, G., Nathan, O., Pentland, A.S.: Decentralizing privacy: using blockchain to protect personal data. In: Proceedings - 2015 IEEE Security and Privacy Workshops, SPW 2015, pp. 180–184. IEEE Computer Society (2015)
51. Chen, Z., Yu, Z., Duan, Z., Hu, K.: Inter-blockchain communication. DEStech Trans. Comput. Sci. Eng. 448–454 (2017)
52. Hitt, M.A., Ireland, R.D., Sirmon, D.G., Trahms, C.A.: Strategic entrepreneurship: creating value for individuals, organizations, and society. Strateg. Entrep. J. **2**, 175–190 (2008)

Author Index